Progress
in Molecular and
Subcellular Biology

10

With Contributions by
P. S. Agutter, K. Dose, H. Stebbings

Edited by
J. Jeanteur, Y. Kuchino,
W. E. G. Müller (Managing Editor),
P. L. Paine

With 18 Figures

Springer-Verlag
Berlin Heidelberg New York
London Paris Tokyo

Professor Dr. WERNER E. G. MÜLLER
Physiologisch-Chemisches Institut
Universität Mainz
Duesbergweg 6
6500 Mainz, FRG

ISBN 3-540-19170-4 Springer-Verlag Berlin Heidelberg New York
ISBN 0-387-19170-4 Springer-Verlag New York Berlin Heidelberg

Library of Congress Catalog Card Number 75-79748

Typesetting: Fotosatz & Design, Berchtesgaden
Offsetprinting and bookbinding: Konrad Triltsch, Graphischer Betrieb, Würzburg
2131/3130-543210

Contents

Contributors

AGUTTER, P. S., Department of Biological Sciences, Napier College, Colinton Road, Edinburgh EH10 5DT, Great Britain

DOSE, K., Institut für Biochemie, Johannes Gutenberg-Universität, Becherweg 30, 6500 Mainz, FRG

STEBBINGS, H., Department of Biological Sciences, University of Exeter, Washington Singer Laboratories, Perry Road, Exeter EX4 1SH, Great Britain

Translocation Along Microtubules in Insect Ovaries

H. Stebbings[1]

A. Introduction

The study of microtubule-associated intracellular translocation is, at present, one of the most exciting areas in cell biology. Most of the research has focussed on the translocation which occurs along nerve axons, but other asymmetric cells, such as chromatophores and various protozoa, have also been studied extensively, and the phenomenon probably occurs to some extent in all cells. It is particularly emphasized too in the ovaries of certain insects, and this is the subject of the present chapter.

B. Insect Ovaries, Their Morphology and Development

Whereas in some orders of insects the oogonia simply develop into oocytes (panoistic ovaries), during oogenesis in other orders the oogonia generate not only oocytes, but also nutritive cells (meroistic ovaries). The way in which this happens has been the subject of much study (see King and Büning 1984) but will not be considered here. Where nutritive cells do occur they show one of two arrangements with respect to the oocytes and it is this aspect which is of particular significance to the topic of intracellular translocation. In some instances the nutritive cells alternate with the oocytes, with which they retain cytoplasmic continuity down the length of the egg tube or ovariole (polytrophic). In other cases, all of the nutritive cells remain confined within an anterior trophic end of the ovary (telotrophic). In the latter, which occur notably in the hemiptera, the nutritive cells are connected with the oocytes, at least during the early previtellogenic stage of oogenesis, by way of cytoplasmic bridges or channels called nutritive tubes. One nutritive tube connects to each oocyte, which means that there may be some 20 nutritive tubes in a single ovariole of some species. Nutritive tubes are usually in the order of 20 μm in diameter and each tube gradually elongates as the oocyte with which it connects passes down the ovariole, so that it may extend many millimetres in some species (Fig. 1).

[1]Department of Biological Sciences, University of Exeter, Washington Singer Laboratories, Perry Road, Exeter, EX4 4QG, Great Britain

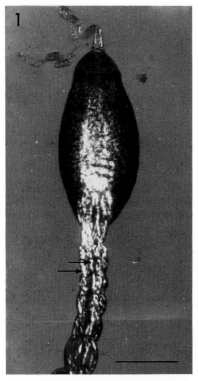

Fig. 1. Anterior portion of ovariole of *Notonecta glauca* viewed in polarized light. The massive array of aligned microtubules within the trophic core and nutritive tubes (*arrows*) renders them birefringent. Bar = 0.4 mm

Nutritive tubes contain massive numbers of aligned microtubules and hence the system has proved, and is continuing to be, extremely valuable for studying microtubule-associated translocation in cells. Moreover, the system of aligned microtubules is a highly dynamic one. Apart from elongating while translocation is occurring and the oocyte is passing down the ovariole, there comes a point at which the oocyte which it supplies starts to accumulate yolk, and in most species at this onset of vitellogenesis the nutritive tube connection is discontinued and the microtubules comprising it disappear.

C. Translocation of Materials Between Nutritive Cells and Oocytes

The concept that materials pass from nutritive cells to the developing oocytes in meroistic insect ovaries is a longstanding one (see Lubbock 1859). Indeed there have been many morphological and histochemical studies of hemipteran telotrophic ovaries which have supported this supposition (see Stebbings 1986). Translocation along nutritive tubes has, however, been confirmed by means of radioactive labelling followed primarily by autoradiography in a number of studies using a variety of insects (see Gutzeit 1986). There have also been reports of the direct observation of movement of components between nutritive cells and oocytes using time-lapse cinematography (Adams and Eide 1972). This has been made considerably more feasible by the development over the past few years of video-enhanced microscopy (see Inoué 1986) and its application to telotrophic ovarioles amongst other systems has allowed the visualization of the movement of particulate material, notably mitochondria, along nutritive tubes (Dittmann et al. 1987).

I. Components Which Translocate Along Nutritive Tubes

Histochemical studies showed the nutritive tubes to be strongly basophylic and autoradiographic data has confirmed that large amounts of RNA synthesized in the nutritive cells pass along the nutritive tubes to the oocytes. Most of the RNA is ribosomal, but mRNA and tRNA also pass (Davenport 1976).

The application of labelled precursors, followed by gel electrophoretic analysis of the contents of nutritive tubes and fluorography has shown label, not only in ribosomal proteins, but in many other polypeptide components of the nutritive tubes (Stebbings et al. 1985). Also, analysis of the polypeptide compositions of nutritive cells, nutritive tubes and oocytes from the same ovary has demonstrated them to be comparable, suggestive of their passage through the system (Sharma and Stebbings 1985). The question of protein transport has also been addressed by introducing fluorescently labelled proteins into the different compartments of the ovary by microinjection, and then monitoring their passage, but the results from this approach are equivocal (Woodruff and Anderson 1984).

With regards the translocation of organelles, apart from microtubules, ribosomes are the most conspicuous components, and in many instances fill the remainder of the nutritive tubes. Such observations, combined with the autoradiographic data, and the fact that nutritive cells have extremely prominent nucleoli, mean that there is little doubt that ribosomes assembled in the nutritive cells pass along the nutritive tubes to the oocytes. Indeed it is a feature of oocytes generally that they accumulate ribosomes during oogenesis. Mitochondria are similarly accumulated by oocytes and in some hemipterans this is accomplished, at least in part, by importation of mitochondria from the nutritive cells via nutritive tubes. The study of a limited number of species suggests that this importation of mitochondria occurs in insects which have rapid oogenesis, but not to any extent in those where oogenesis is protracted (Hyams and Stebbings 1977).

II. Characterization of Movement

Estimations of the rates of movement of RNA down nutritive tubes, from autoradiographs show movement to be slow. In *Notonecta* (Fig. 2) this was assessed at 0.5 mm/day (Macgregor and Stebbings 1970), and a similar rate was recorded in the same way for *Pyrrhocoris* (Mays 1972). Interestingly, in the latter case a much faster component (200 μm/h) superimposed on the slower was also recorded.

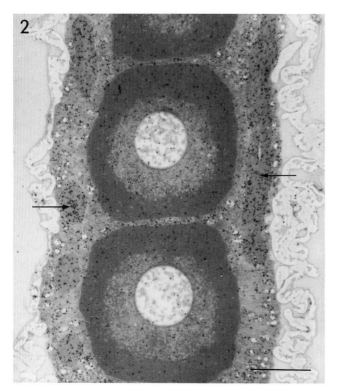

Fig. 2. Autoradiograph of a longitudinal section through a row of oocytes within an ovariole. Twenty-four h after the application of ^{3}H-uridine there is heavy labelling over the nutritive tubes (*arrows*) indicating the presence of newly synthesized RNA. Bar = 40 μm (MacGregor and Stebbings 1970)

The rates of movement of mitochondria along nutritive tubes appear to be considerably greater than the bulk movement of RNA. From direct observations these have been estimated at approximately 3 μm/min in *Dysdercus* (Dittmann et al., 1987) or comparable to the faster movement recorded by autoradiography. Surprisingly perhaps, the movement of mitochondria was seen in the latter report to be bidirectional.

Evidence of different rates of movement along individual nutritive tubes clearly argues against a cohesive flow of all components along these channels. This is supported by the fact that different polypeptide components of nutritive tubes incorporate labelled precursors at different rates and independently of the proportions in which they exist in the tubes (Stebbings et al. 1985). By the same token, selectivity over what translocates also appears to exist, and suggestions that the entire contents of the nutritive cells pass down nutritive tubes to the oocytes are patently incorrect. Indeed it has already been discussed that mitochondria translocate in some species, but not in others, and plainly in no cases do the nutritive cell nuclei pass to the haploid gametes.

D. Ultrastructure of Nutritive Tubes

Electron microscopy of nutritive tubes in telotrophic ovaries of all hemipterans so far studied has shown them to be packed with longitudinally orientated microtubules (Brunt 1970; Huebner and Anderson 1970; Macgregor and Stebbings 1970). This is not the case, however, for the much less extensive nutritive tubes of polyphagous coleopterans (Büning 1979; Stebbings 1981), where microtubules do occur but not in such large numbers or any conspicuous orientation. In hemipteran nutritive tubes the microtubule system may be very extensive. In *Notonecta*, for example, a transverse section of a single tube shows in the order of 30,000 microtubule profiles.

The dynamics of the microtubule system match those of the nutritive tubes they comprise. This has been illustrated by investigations into the way the microtubule system is assembled and elaborated in *Dysdercus* (Hyams and Stebbings 1979 a) as the nutritive tubes themselves widen and lengthen. These studies have also shown that the microtubules within a nutritive tube adopt an altered arrangement and rapidly depolymerize when the nutritive cell contribution to an oocyte ceases and the nutritive tube becomes redundant in late oogenesis.

In transverse sections of functional nutritive tubes, the microtubules can be seen to be evenly spaced, and while the spacing is the same in all of the functional tubes of a particular species, the spacing varies considerably from species to species (Hyams and Stebbings 1977). Interestingly too, the separation of the microtubules in the limited number of species looked at shows a correlation with the size of components which pass between them along the tubes. In *Corixa*, for example, where the microtubules are very closely packed, mitochondria which are larger than the distances between adjacent microtubules do not pass down the tubes. In *Oncopeltus*, on the other hand, the microtubules show a much greater separation and mitochondria do pass down the nutritive tubes of this species (cf. Figs. 3 and 4).

An obvious feature of nutritive tube microtubules is that they each appear surrounded by an electron-clear zone into which other organelles do not encroach. The basis behind the separation is not known for certain, and while one view is that the zone is maintained by invisible microtubule-associated proteinaceous structure (Amos 1979), there is some evidence that it results from electrostatic repulsion between adjacent microtubules, and microtubules and surrounding organelles (Stebbings and Hunt 1982). Moreover, while nutritive tube microtubules do possess microtubule-associated proteins (MAPs) (Stebbings et al. 1986) distinct regular projections have not been observed from nutritive tube microtubules by any technique used so far, although linkages between microtubules and translocating mitochondria cannot be ruled out (Stebbings and Hunt, 1987).

As well as being parallel, it appears that the microtubules which pack nutritive tubes are all of a single polarity. This has been shown (Stebbings and Hunt 1983) using the tubulin hook-decoration technique which has been applied to· a number of microtubule systems. The technique also reveals that the plus or fast growing ends of the microtubules occur at the ends of the nutritive tubes adjacent to the trophic region,

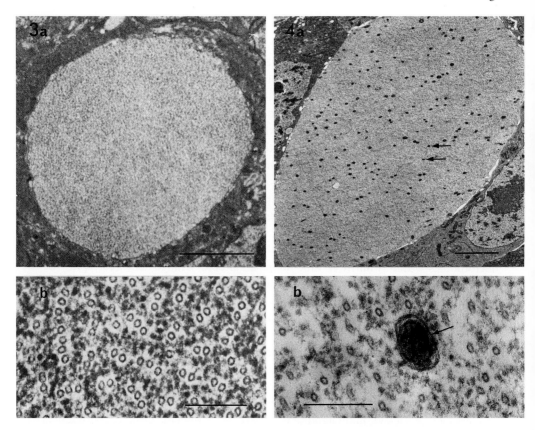

Figs. 3. and 4. (*a*) Low power and (*b*) high power electron micrographs of transverse sections through ovarioles of *Corixa punctata (Fig. 3)* and *Oncopeltus fasciatus (Fig. 4)*. The presence of mitochondria (*arrows*) in the nutritive tubes of *Oncopeltus* but not *Corixa* correlates with the lower density of microtubules found in the former species. Figs *3(a)* and *4(a)* bar = 3 μm, Figs *3(b)* and *4(b)* bar = 0.2 μm

with the minus, or slow growing ends close to the oocytes. No microtubule organizing centres have yet been identified in the ovarioles.

Microtubules are the only linear elements seen in hemipteran nutritive tubes and no filaments are present. Microfilaments have, however, been observed around the nutritive region from which the nutritive tubes emanate (Huebner and Gutzeit 1986).

E. Mechanisms of Microtubule-Associated Transport Along Nutritive Tubes

An appreciation of the nature of the movement along nutritive tubes is clearly fundamental to understanding the underlying mechanisms involved. Accumulating evidence suggests that movement along nutritive tubes occurs relative to, rather than in conjunction with, the microtubules in the tube. First, labelling studies of newly

synthesized polypeptides into nutritive tubes have shown tubulin, the major component of the microtubules, to be much less heavily labelled in relation to almost all
other proteins. Secondly, the direct observations of mitochondrial movements show
that these must occur independently of microtubule movement.

Materials therefore appear to pass down nutritive tubes between the microtubules,
and indeed, as mentioned in a previous section, the spacings of the microtubules
could conceivably exert some selectivity over what is transported. Apart from this,
the extensive microtubule system within the tubes may be cytoskeletal, involved simply in the development and maintenance of the transport channel.

The recording of different rates of movement is also most significant, and argues
against a cohesive mass flow along nutritive tubes. Were the latter a possibility, translocation could have been accounted for solely by a synthesis at one end of the system
followed by displacement away from the site of synthesis. Multiple rates of transport,
however, suggest a greater complexity, and while synthetic pressure may explain the
movement of one component down the nutritive tubes, additional mechanisms must
also be considered.

Another apparent possibility is that components might pass along nutritive tubes by
electrophoresis. This suggestion is based on the fact that electrophysiological measurements have shown there to be a small potential gradient along the nutritive tubes
with the oocyte end positive relative to the nutritive cell end (Dittmann et al. 1981).
The idea seemed to be supported to an extent by the tracking of microinjected proteins which migrated according to their charge. More recent evaluation of the possibility has, however, demonstrated that the potential difference which exists between nutritive cells and oocytes would be too small to establish an effective electrophoretic
gradient along the nutritive tubes in telotrophic ovarioles (Woodruff and Anderson
1984).

Recently attention has turned once again to the possibility of an active role for microtubules in translocation along nutritive tubes prompted mainly by exciting observations of translocation along microtubules isolated from neuronal axoplasm (Allen
et al. 1985; Schnapp et al. 1985). Since microtubules were first described in cells, they
have been linked with translocation of a range of components in a wide variety of cell
types. The most extensively studied have been the movement of vesicles along
neuronal axons, the movement of pigment granules in chromatophores of various animals, the translocation of cytoplasm in highly asymmetric protozoa and the subject of
this review, the transport of various components down nutritive tubes. In each case
the movement occurs along an aligned system of microtubules, but the question of
whether the microtubules cause or merely facilitate the movement seemed an intracti-
ble one. Experiments with antimitotic drugs, which when applied to some of the systems destroyed the microtubules and at the same time arrested movement, only
served to show that microtubules were essential for movement, but did not discrimi-
nate between their having an active or a passive role in the movement.

. The breakthrough to answering this question came with the development, in the
early 1980s, of video-enhanced contrast microscopy (Allen et al. 1981; Inoué and Til-

ney 1982). This immediately allowed the visualization in living preparations of cellular components considerably smaller than the limit of resolution of the light microscope which is used to view them and microtubules became visible not only in living cells (Hayden etal. 1983) but also in isolated cytoplasmic slurries (Allen and Allen 1983). The importance of being able to observe living microtubules for the first time was that it allowed their involvement in cellular dynamics to be investigated directly. Indeed, on application of ATP to preparations of isolated squid axoplasm, the most popular material for studying the role of microtubules in translocation, vesicles were seen to translocate along the microtubules in a manner comparable to that seen in intact axons. This therefore constituted the first direct evidence that microtubules can be actively involved in translocation along their length (Allen etal. 1985; Vale etal. 1985c), and resolved a long-standing dilemma. The movements seen in vitro along axoplasmic microtubules have now been carefully documented (see Weiss 1986) and these included not only the translocation of particles along microtubules, but surprisingly the movement of the microtubules themselves relative to the glass substrate.

The main problem with using axoplasm for such studies is that it contains a variety of cytoskeletal components and while care was taken to ensure that the linear elements seen were indeed single microtubules (Schnapp etal. 1985) the presence of other components such as microfilaments does introduce a degree of complexity to the preparations. Various approaches have been adopted to circumvent this, mainly by using reconstituted models and in one example axoplasmic vesicles have been shown to translocate along flagellar microtubules (Gilbert etal. 1985). Translocation has also been shown to occur along microtubules which have assembled from microtubule-organizing centres (Vale etal. 1985b).

The demonstration of translocation along microtubules using video-enhanced contrast microscopy has not been totally confined to axoplasmic preparations and neuronal components, since a similar movement has subsequently been demonstrated along the extracted cytoskeleton of a giant freshwater amoeba (Koonce etal. 1986). Again such preparations contain both microtubules and microfilaments, which have to be separated experimentally.

Nutritive tubes, by contrast, present an example where translocation occurs solely in conjunction with microtubules. Indeed, the numbers of microtubules involved are substantial. They are so great as to render the nutritive tubes strongly birefringent, and fortuitously this feature has allowed a technique to be developed for microdissecting nutritive tubes from ovarioles (Hyams and Stebbings 1979b). In this way the intact nutritive tube translocation channels can be isolated for morphological, and also biochemical studies (Fig. 5).

Nutritive tube preparations have been examined using differential interference contrast microscopy with an element of video enhancement. Moreover, on addition of ATP to isolated nutritive tubes of *Oncopeltus,* mitochondria which are found in considerable numbers in the nutritive tubes of this species are seen to translocate along linear elements which emerge from the sides and the frayed ends of the isolated tubes (Stebbings and Hunt, 1987). These linear elements have been shown by electron

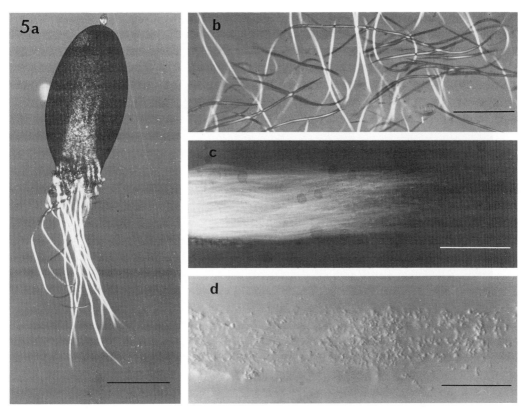

Fig. 5. In polarized light it is possible to *(a)* microdissect an ovariole to isolate the nutritive tubes which *(b)* can then be pooled in a microtubule-stabilizing buffer. At higher power *(c)* the nutritive tube appears wispy but with differential interference contrast miroscopy *(d)* it can be seen to extend beyond the wispy birefringence, and particles within the nutritive tube are visible. *(a)* and *(b)* bar = 0.4 mm, *(c)* and *(d)* bar = 40 μm

microscopy to consist of bundles of parallel microtubules. From video images the mitochondria appear attached to the microtubules while translocating (Fig. 6), and electron microscopy has also shown that mitochondria have an association with the microtubules, while the ribosomes, which also translocate along the nutritive tubes, do not.

The movements of the mitochondria along the isolated nutritive tube microtubules are comparable to those of vesicles along isolated axoplasmic microtubules, both in terms of manner and rate. They are considerably faster than those reported for mitochondria in nutritive tubes in vivo (Dittmann et al. 1987), but this could be explained by the fact that the isolated preparations have been freed from the logjamming effect of confinement within the nutritive tubes, or the greater availability of ATP, or both. Indeed, no movement was observed in the isolated nutritive tubes themselves.

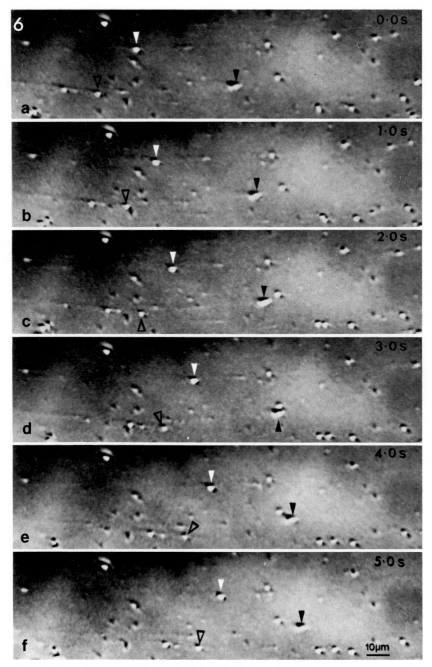

Fig. 6. Stills taken at 1-s intervals from the video record of a nutritive tube of *Oncopeltus fasciatus*, observed using differential interference contrast microscopy. Many mitochondria (see *arrows*) have moved along visible tracks in the course of observations

Such observations demonstrate that, as in other systems the microtubules in nutritive tubes are capable of actively translocating components along their length. They do not preclude other possibilities and it is conceivable that movements of different components at different rates down nutritive tubes have different bases. At present it would seem a faster microtubule-based translocation of mitochondria might be superimposed on the movement of a slower component, perhaps driven by synthesis, and in many ways such a system is comparable to the movement which occurs down nerve axons.

F. The Motor(s) for Microtubule-Based Translocation

Following the discovery that translocation could be observed along neuronal microtubules on addition of ATP, came the search for the microtubule motor(s). Again using squid, experiments showed that motility was promoted using a factor from axoplasm. The key to the identification of the motor, believed to be an ATPase, proved to be the addition of the non-hydrolyzable ATP analogue, AMP-PNP which resulted in the tight binding of vesicles to microtubules to form a necklace pattern (Lasek and Brady 1985) and the concomitant binding to microtubules of a polypeptide which possessed the predicted characteristics of the translocation motor (Brady 1985; Vale et al. 1985 a). This polypeptide was then purified, shown to promote translocation and named kinesin. It has subsequently been further characterized in terms of the translocation which it promotes (Vale et al. 1985 b) and the ATPase activity it possesses (Kuznetsov and Gelfand 1986), and evidence is accumulating to link one with the other (Cohn et al. 1987).

Kinesin has now been looked for and identified in a variety of neuronal and non-neuronal sources. With regards the intracellular translocation of components, however, even in axons kinesin can only be part of the answer, in that while it may be responsible for anterograde movement, retrograde movement also occurs in such axons and must be accounted for. It seems most likely in fact that different translocators and multiple translocators may exist in different cellular situations and that these may act in concert.

An increasing number of microtubule-associated proteins, some with ATP-sensitive binding to microtubules and some with demonstrated ATPase activity, have been identified, again from a variety of neuronal and non-neuronal sources (Collins and Vallee 1986; Gilbert and Sloboda 1986; Hollenbeck and Chapman 1986; Koszka et al. 1987). Certain of these have the properties one might expect of a microtubule-based translocator, although the ability to promote translocation remains to be shown in many instances.

Telotrophic insect ovaries, and more specifically the nutritive tube translocation channels lend themselves to similar studies. The entire polypeptide composition of the nutritive tubes (presumably including any microtubule-based translocators) has already been determined for one species, *Notonecta* (Hyams and Stebbings 1979 b;

Sharma and Stebbings 1985; Stebbings et al. 1986). For the future, the extreme simplicity of the system, the fact that it can be isolated intact and reactivated, and the advantage that microtubules can be isolated from it by a wide range of approaches, render it eminently suitable for studying microtubule-based translocation. Importantly too, its study serves to broaden investigations into what is almost certainly a general cellular phenomenon.

Acknowledgements. Research carried out in my laboratory, and discussed in this chapter, has been supported by funds from the Science and Engineering Research Council (U.K.) and the University of Exeter Research Fund.

References

Adams TS, Eide PE (1972) A method for the in vitro stimulation of house fly egg development with a juvenile analog. Gen Comp Endocrinol 18:12–21

Allen RD, Allen NS (1983) Video-enhanced microscopy with a computer frame memory. J Microsc 129:3–17

Allen RD, Allen NS, Travis JL (1981) Video-enhanced contrast, differential interference contrast (AVEC-DIC) microscopy: a new method capable of analyzing microtubule-related motility in the reticulopodial network of *Allogromia laticollaris*. Cell Motil 1:291–302

Allen RD, Weiss DG, Hayden JH, Brown DT, Fujiwake H, Simpson M (1985) Gliding movement of and bidirectional transport along single native microtubules from squid axoplasm: evidence for an active role of microtubules in cytoplasmic transport. J Cell Biol 100:1736–1752

Amos LA (1979) Structure of microtubules. In: Roberts K, Hyams JS (eds) Microtubules. Academic Press, London

Brady ST (1985) A novel brain ATPase with properties expected for the fast axonal transport motor. Nature 317:73–75

Brunt AM (1970) Extensive system of microtubules in the ovariole of *Dysdercus fasciatus* Signoret (Heteroptera: Pyrrhocoridae). Nature 228:80–81

Büning J (1979) The telotrophic nature of ovarioles of polyphage Coleoptera. Zoomorphologie 93:51–57

Cohn SA, Ingold AL, Scholey JM (1987) Correlation between the ATPase and microtubule translocating activities of sea urchin egg kinesin. Nature 328:160–163

Collins CA, Vallee RB (1986) Characterization of the sea-urchin egg microtubule-activated ATPase. J Cell Sci Supp 5:197–204

Davenport R (1976) Transport of ribosomal RNA into the oocytes of the milkweed bug, *Oncopeltus fasciatus*. J Insect Physiol 22:925–926

Dittmann F, Ehni R, Engels W (1981) Bioelectric aspects of the hemipteran telotrophic ovariole (*Dysdercus intermedius*). Roux's Arch Dev Biol 190:221–225

Dittmann F, Weiss DG, Münz A (1987) Movement of mitochondria in the ovarian trophic cord of *Dysdercus intermedius* (Heteroptera) resembles nerve axonal transport. Roux's Arch Dev Biol 196:401–413

Gilbert SP, Allen RD, Sloboda RD (1985) Translocation of vesicles from squid axoplasm on flagellar microtubules. Nature 315:245–248

Gilbert SP, Sloboda RD (1986) Identification of a MAP 2-like ATP-binding protein associated with axoplasmic vesicles that translocate on isolated microtubules. J Cell Biol 103:947–956

Gutzeit HO (1986) Transport of molecules and organelles in meroistic ovarioles of insects. Differentiation 31:155–165

Hayden JH, Allen RD, Goldman RD (1983) Cytoplasmic transport in keratocytes: direct visualization of particle translocation along microtubules. Cell Motil 3:1–19

Hollenbeck PJ, Chapman K (1986) A novel microtubule-associated protein from mammalian nerve shows ATP-sensitive binding to microtubules. J Cell Biol 103:1539–1545

Huebner E, Anderson E (1970) The effects of vinblastine sulfate on the microtubular organization of the ovary of *Rhodnius prolixus*. J Cell Biol 46:191–198

Huebner E, Gutzeit H (1986) Nurse cell-oocyte interaction: a new F-actin mesh associated with the microtubule-rich core of an insect ovariole. Tissue Cell 18:753–764

Hyams JS, Stebbings H (1977) The distribution and function of microtubules in nutritive tubes. Tissue Cell 9:537–545

Hyams JS, Stebbings H (1979 a) The formation and breakdown of nutritive tubes – massive microtubular organelles associated with cytoplasmic transport. J Ultrastruct Res 68:46–57

Hyams JS, Stebbings H (1979 b) The mechanism of microtubule associated cytoplasmic transport. Isolation and preliminary characterisation of a microtubule transport system. Cell Tissue Res 196:103–116

Inoué S (1986) Video microscopy. Plenum, New York

Inoué S, Tilney LG (1982) Acrosomal reaction of *Thyone* sperm. I. Changes in the sperm head visualized by high resolution video microscopy. J Cell Biol 93:812–819

King RC, Büning J (1984) The origin and functioning of insect oocytes and nurse cells. Comp Insect Physiol Biochem Pharmacol 1:37–82

Koonce MP, Euteneuer U, Schliwa M (1986) *Reticulomyxa:* a new model system of intracellular transport. J Cell Sci Suppl 5:145–159

Koszka C, Foisner R, Seyfert HM, Wiche G (1987) Isolation of a Ca^{2+}-protease resistant high Mr microtubule binding protein from mammalian brain: characterization of properties partially expected for a dynein-like molecule. Protoplasma 138:54–61

Kuznetsov SA, Gelfand VI (1986) Bovine brain kinesin is a microtubule-activated ATPase. Proc Natl Acad Sci USA 83:8530–8534

Lasek RJ, Brady ST (1985) Attachment of transported vesicles to microtubules in axoplasm is facilitated by AMP-PNP. Nature 316:645–647

Lubbock J (1859) On the ova and pseudova of insects. Philos Trans R Soc Lond 149:341–369

Macgregor HC, Stebbings H (1970) A massive system of microtubules associated with cytoplasmic movement in telotrophic ovarioles. J Cell Sci 6:431–449

Mays U (1972) Stofftransport im Ovar von *Pyrrhocoris apterus* L. Z Zellforsch 123:395–410

Schnapp BJ, Vale RD, Sheetz MP, Reese TS (1985) Single microtubules from squid axoplasm support bidirectional movement of organelles. Cell 40:455–462

Sharma KK, Stebbings H (1985) Electrophoretic characterization of an extensive microtubule-associated transport system linking nutritive cells and oocytes in the telotrophic ovarioles of *Notonecta glauca*. Cell Tissue Res 242:383–389

Stebbings H (1981) Observations on cytoplasmic transport along ovarian nutritive tubes of polyphagous coleopterans. Cell Tissue Res 220:153–161

Stebbings H (1986) Cytoplasmic transport and microtubules in telotrophic ovarioles of hemipteran insects. Int Rev Cytol 101:101–123

Stebbings H, Hunt C (1982) The nature of the clear zone around microtubules. Cell Tissue Res 227:609–617

Stebbings H, Hunt C (1983) Microtubule polarity in the nutritive tubes of insect ovarioles. Cell Tissue Res 233:133–141

Stebbings H, Hunt C (1987) The translocation of mitochondria along insect ovarian microtubules from isolated nutritive tubes: a simple reactivated model. J Cell Sci 88:641–648

Stebbings H, Sharma K, Hunt C (1985) Protein turnover in the cytoplasmic transport system within an insect ovary — a clue to the mechanism of microtubule-associated transport. FEBS Lett 193:22–26

Stebbings H, Sharma KK, Hunt C (1986) Microtubule-associated proteins in the ovaries of hemipteran insects and their association with the microtubule transport system linking nutritive cells and oocytes. Eur J Cell Biol 42:135–139

Vale RD, Reese TS, Sheetz MP (1985a) Identification of a novel force-generating protein, kinesin, involved in microtubule-based motility. Cell 42:39–50

Vale RD, Schnapp BJ, Mitchison T, Steuer E, Reese TS, Sheetz MP (1985b) Different axoplasmic proteins generate movement in opposite directions along microtubules in vitro. Cell 43:623–632

Vale RD, Schnapp BJ, Reese TS, Sheetz MP (1985c) Movement of organelles along filaments dissociated from the axoplasm of the squid giant axon. Cell 40:449–454

Weiss DG (1986) Visualization of the living cytoskeleton by video-enhanced microscopy and digital image processing. J Cell Sci Suppl 5:1–15

Woodruff RI, Anderson KL (1984) Nutritive cord connection and dye-coupling of the follicular epithelium to the growing oocytes in the telotrophic ovarioles in *Oncopeltus fasciatus*, the milkweed bug. Roux's Arch Dev Biol 193:158–163

Nucleo-Cytoplasmic Transport of mRNA: Its Relationship to RNA Metabolism, Subcellular Structures and Other Nucleocytoplasmic Exchanges

Paul S. Agutter[1]

A. Introduction

I. Scope of This Review

The exchange of materials between the two major components of a eukaryotic cell is a generally interesting study. It bears directly on a number of topical research problems, such as the mechanisms of histone accumulation and nucleosome assembly in nuclei, the modulation of intranuclear processes by cytoplasmic regulators, and the specific, selective transport of mature ribonucleoproteins to the cytoplasm. Less directly, it has implications for general issues such as the nature and significance of intracellular skeletal structures, the regulation of gene expression and the mechanisms of cellular differentiation. Techniques from a variety of disciplines have been brought to bear on the study of nucleocytoplasmic transport: ultrastructural and biochemical characterizations of the nuclear envelope and of other macromolecular assemblies, microinjection, heterokaryon and nuclear transplantation experiments, and experiments with isolated nuclei have all proved important. With the aid of such approaches, advances have been made during the last few years, and the field is ripe for reviewing.

However, while all this is clear from a cursory glance at the literature, it is less evident why a reviewer should focus on just one aspect of nucleocytoplasmic exchanges, viz. messenger RNA transport. Essentially, the reason is that mRNA transport seems to be a "solid-state" process, i.e. the transported material is structurally bound rather than freely soluble before, during and after movement across the nuclear envelope. There is no compelling reason to believe that other nucleocytoplasmic exchanges are of the same kind. This implies two things: first, reasonable mechanistic models of mRNA transport and of other nucleocytoplasmic exchanges are likely to be very different; and second, mRNA transport and other such exchanges are likely to be amenable to study by quite different methods. (The evidence for this position is discussed in Sect. A.III, below.) Moreover, hnRNA processing, another field that de-

[1]Department of Biological Sciences, Napier College, Colinton Road, Edinburgh EH10 5DT, Great Britain

mands a reviewer's attention because of recent advances, is normally a prerequisite for mRNA transport and needs to be discussed along with it, but it is not directly relevant to other nucleocytoplasmic transport processes. The same applies to advances in our knowledge of ribonucleoprotein (RNP) organization and its structural attachments in nucleus and cytoplasm. For these reasons, this review focuses on progress in the analysis of mRNA transport and its relationship to post-transcriptional processing and to intracellular structures.

Nevertheless, it would be inappropriate to exclude discussion of other nucleocytoplasmic transport processes altogether, for at least the following reasons. First, recent progress in the study of such processes is inherently interesting and important. Second, protein transport is at least indirectly relevant, because cytoplasmic proteins that regulate nucleocytoplasmic mRNA transport have to cross the nuclear envelope, and because messenger-associated proteins might have a mechanstic role in taking the RNP through the pore-complexes. Third, although the "solid-state" view of mRNA transport is now widely accepted, it is not unchallengeable. If it proved to be wrong, then the argument in the preceding paragraph would have been overstated. Fourth, any individual study of nucelocytoplasmic exchanges can potentially increase the array of methods available to the field as a whole, despite the apparent distinctiveness of mRNA transport. A review that was too narrowly focussed could encourage its readers to ignore parallel advances that might, albeit indirectly, enrich their own studies.

Several existing reviews are relevant to the range of fields mentioned so far. De Robertis (1983) has reviewed in situ nucleocytoplasmic transport studies concisely and clearly. I have reviewed the literature pertinent to messenger transport, with emphasis on in vitro studies, previously (Agutter, 1984; Agutter and Thomson 1984), and the articles in Volume 9 of Busch's series *The Cell Nucleus* should be read. On the nuclear envelope, the authoritative early reviews of Franke (1974) and Franke and Scheer (1974) remain valuable, and the discussion of pore-complex ultrastructure by Maul (1977) has not yet been surpassed. Advances in RNA processing have outstripped all reviews extant at the time of writing this article, but the discussion by Nevins (1983) remains valuable, and the relationship between HnRNA processing and HnRNP structure is authoritatively discussed by Pederson (1983). On the cytoskeleton, the articles in Volume 99, Part 1, No. 2 of *J Cell Biol* give the reader a grasp of current view and approaches. With regard to the nuclear matrix controversy, it is instructive to compare, for example, the reviews by Berezney (1979) and Hancock and Boulikas (1982). I shall mention other relevant reviews later in the text.

The present review is structured as follows. Section A.II explains the nomenclature used in the rest of the article, and Section A.III presents the two opposing perspectives from which nucleocytoplasmic transport in general, and mRNA transport in particular, have been studied. In Section B, I review the three types of study that are most directly relevant to nucleocytoplasmic transport: studies on the nuclear envelope, in situ studies using e. g. microinjection and heterokaryon techniques, and in vitro studies using isolated nuclei or resealed nuclear envelope vesicles. I evaluate the contributions of these studies to our knowledge of nucleocytoplasmic transport, and in particular to mRNA transport. Section C is devoted to advances in our understanding of the metabolism and organization of mRNA and its precursors in vivo, and contains a discussion of the controversy surrounding the nuclear matrix concept. In Section D, I review current progress in our understanding of the mechanism of mRNA transport itself. Section E is devoted to our knowledge of how mRNA transport is regulated, and to the biological implications of its regulation.

II. Nomenclature

As far as possible, I shall use the terms that are most widely employed by workers in this and related fields.

The structures that I shall discuss include various RNP complexes, the cytoskeleton, the nuclear matrix and the nuclear envelope itself. The relevant RNP complexes are (a) nucleus-restricted messenger precursors bound to a specific set of polypeptides (HnRNP), and (b) cytoplasmic messengers, also associated with a more or less specific set of polypeptides (mRNP). It is now generally agreed that the cytoskeleton comprises three kinds of structures: microtubules, microfilaments and intermediate filaments. A fourth, the "microtrabecular lattice", might need to be added (Wolosewick and Porter 1976). At present, it seems that amongst these elements, the intermediate filaments are most directly relevant to mRNA transport. The nuclear matrix, if it exists at all, bears the same general relationship to the nucleoplasm as the cytoskeleton does to the cytoplasm. It is not nearly so well characterized, and although most workers in the field accept or assume its reality, some authorities regard it as an artefact. The nuclear envelope comprises four ultrastructurally (and probably biochemically) distinct components: the outer and inner nuclear membranes, the lamina, and the pore-complexes. Nuclear envelope architecture will be discussed in more detail in Section B.I.

The most important function to be discussed in this review is mRNA transport itself. If this is a solid-state process, then it must, minimally, comprise three stages: detatchment of mature mRNA or mRNP from intranuclear binding sites; passage through the pore complexes of the nuclear envelope; and association with insoluble structures in the cytoplasm. I refer to these three stages, respectively, as *release, translocation* and *cytoskeletal binding*. When in vitro models, involving incubation of isolated nuclei, are used to investigate mRNA transport, the phenomenon being studied is logically distinct from mRNA transport in vivo, because the cytoskeletal binding stage is missing. I use the term *mRNA efflux* to describe the appearance of messenger in the incubation supernatant in such experiments, reserving the term *mRNA transport* for the in vivo export process. The other important function to be discussed, HnRNA processing, comprises three main stages: methylation of the 5' end of the RNA molecule (capping), the attachment of 150–200 adenylic acid residues to the 3' end (adenylation), and the excision of intron transcripts (splicing). Other events, such as 3' terminal cleavage and endomethylations, will be mentioned in passing: their relevance to mRNA transport and metabolism is not yet clear.

This covers the main items of terminology in this article. Other terms that are introduced during the discussion will be defined where appropriate.

III. Perspectives on Messenger RNA Transport

Figure 1 gives schematic representations of three possible modes of nucleocytoplasmic exchange. In Fig. 1a, the transport substrate is soluble and freely diffusible in both nucleoplasm and cytoplasm. Exchange is essentially by diffusion through the pore-complexes, though interaction of intramolecular signals with pore-complex components, and facilitated or energy-dependent movement in and through the pore-complex itself, are not excluded. In Fig. 1b, the same principles obtain. The difference is that in either cytoplasm, or nucleoplasm, or both, the transport substrate is biphasically distributed. It is (potentially) in steady state between aqueous and solid (structural) phases. It seems that Fig. 1b is generally a more realistic perspective than Fig. 1a; for instance, karyophilic proteins (i.e. those with very high nucleus/cytoplasm concentration ratios) seem to be karyophilic largely because of their high affinity for intranuclear structures (see Sect. B.II). Nevertheless, the distinction between Fig. 1a and 1b is not fundamental. In both cases, it is accepted that the transport substrate is in aqueous solution at the moment of entering and the moment of leaving the pore-complex.

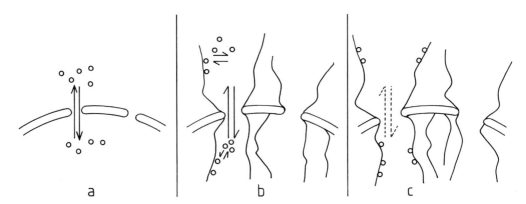

a   b   c

Fig. 1. Three schematic representations of nucleocytoplasmic transport processes. *(a)* and *(b)* represent versions of the solution-diffusion perspective, and *(c)* represents the solid-state perspective. The fundamental point is that in *(c)*, the transported material (represented by *small open circles*) always remains bound to the fibils running to and from the pore-complexes. In *(b)*, the material exchanges between bound and soluble phases. In both *(a)* and *(b)*, it is the soluble material that approaches and leaves the pore-complex, though binding to "carrier" sites within the pore might occur

The scheme shown in Fig. 1c *is* fundamentally distinct. In this case, the transport substrate is bound to structures, themselves contiguous or continuous with the pore-complex, both when it enters the pore and when it leaves. There is no element of diffusion. In a previous discussion of such schemes (Agutter 1985a), I used the terms *solution-diffusion perspective* to describe Fig. 1a, and *solid-state perspective* to describe Fig. 1c. I shall use the same terms here, noting only that I intend Fig. 1b to be considered an alternative, probably more realistic, representation of the solution-diffusion viewpoint. In this section, I shall review the reasons for believing that although Fig. 1b (or Fig. 1a) adequately represents most nucleocytoplasmic exchanges, only Fig. 1c adequately represents mRNA transport.

Generally, it has been clear for some time that the mechanisms underpinning nucleo-cytoplasmic mRNA distribution are radically different from those underpinning, say, nucleocytoplasmic protein distribution, and evidence to this effect continues to accumulate. For instance, Rao and Prescott (1967) showed that the distribution of mRNA, unlike that of proteins, is not detectably altered during mitosis. This implies that although protein distribution depends to some extent on nuclear envelope integrity, RNA distribution does not. Drummond et al. (1985) found that mRNA microinjected into at least some parts of the cytoplasm of *Xenopus* oocytes remained locally immobilized. Sugawa and Uchida (1985) have obtained an antibody, apparently against a 55 kDa intranuclear protein, that blocks mRNA transport in situ but has no effect on protein transport. The fact that nuclei show more or less normal RNA restriction during the best-studied in vitro mRNA efflux experiments (see Sect. B.III), while proteins are rapidly lost from isolated nuclei (Paine et al. 1983), is also significant. Moreover, it is obvious that molecular size is not an important determinant of nucleocytoplasmic RNA distribution; for proteins, size seems to be the most generally important determinant, though not the only one (Paine 1985; see also Sect. B.II). The accumulation of this kind of evidence strongly suggests that mRNA transport is mechanistically distinctive, and that mRNA and its precursors in situ are more or less always immobile. This general evidence is corroborated by some more specific findings.

Gross mechanical damage to the *Xenopus* oocyte nuclear envelope does not alter the nucleocytoplasmic RNA distribution, suggesting that both mRNP and, more significantly, HnRNP particles cannot escape from their usual compartments by diffusion (cf. Feldherr 1980). In contrast, proteins (except for highly karyophilic proteins, and intermediate filament components and similar structural units) are redistributed more or less immediately following such damage (Feldherr and Pomeratz 1978; Feldherr and Ogburn 1980). Newly transcribed RNA is bound to a nuclear protein fibrillar structure (Jackson et al. 1981; Jackson and Cook 1985), and it seems that transcription complexes are associated with the fibrils. This structure, the "nuclear cage", might be similar to the putative nuclear matrix. In any case, a large number of different "nuclear matrix" preparations have now been described that have essentially all the HnRNA bound to them (Herman et al. 1978; Miller et al. 1978; Agutter and Birchall 1979; Long et al. 1979; Berezney 1980; Van Eekelen and Van Venrooij 1981).

However these results are to be interpreted, they strongly suggest the assembly of HnRNP in vivo into high-order structures: many isolated "matrices" disintegrate on ribonuclease treatment (Kaufmann et al. 1981) and the view is widely held that not only nascent transcripts, but all HnRNA, is matrix-bound. The fact that splicing, even if it involves the removal of only one small intron, seems to be a prerequisite for the export of most mRNAs to the cytoplasm in normal cells (Gruss et al. 1979), emphasizes the dependence of RNA transportability on properties other than molecular size. This fact is more readily accommodated by the solid-state perspective than by its solution-diffusion alternative. In terms of the former viewpoint, a reasonable hypothesis would be that introns prevent release, e.g. by direct attachment to the matrix; release is therefore consequent on the last splicing step. In terms of the latter viewpoint, the intron could, presumably, only act in a way analogous to a "stop" signal in integral protein insertion into membranes: the implicit claim about receptors for such "stops" in the pore-complex would be difficult to test experimentally.

These results and interpretations very strongly suggest that HnRNA is always in a solid phase in vivo, but it should be emphasized that the nature of the solid-phase structure remains controversial. Translocation across the nuclear envelope is a transient process, and stable attachments between messenger molecules and pore-complexes would not be expected. Nevertheless, the existence of high-affinity RNA binding sites [showing some specificity for poly(A)] in the nuclear envelope suggests that messengers are most unlikely to be in a diffusible state during this stage of transport to the cytoplasm (McDonald and Agutter 1980; Bernd et al. 1982a). Thus, both within the nucleus and at the envelope, it seems highly probable that the RNA is in a solid, not an aqueous, phase. Finally, so far as the cytoplasm is concerned, there is abundant evidence for the association of mRNP with cytoskeletal elements in vivo. The main aspects of this are as follows.

Cultured cells from which the membranes and soluble components have been removed by non-ionic detergent treatment contain the polysomes that were actively translated in vivo, and these cytoskeleton-bound polysomes can be translated to give normal products when tRNAs, amino acids, the appropriate soluble factors and GTP are added to the extracted cells. Non-translatable messengers are extracted by the detergent (Van Venrooij et al. 1981). Cervera et al. (1981) reported that polysomes associated with the rough endoplasmic reticulum are in fact bound to protein fibrils continuous with the cytoskeleton. That cytoskeletal binding is a precondition for translation has been reported from other laboratories (Moon et al. 1982; Moon et al. 1983; Bonneau et al. 1985). The in situ hybridization studies by Jeffery (1982) are particularly interesting, because they demonstrate the apparently complete association of newly transported mRNA with the cytoskeleton. Bag (1984) showed that cytoskeleton-bound messengers exchanged their associated proteins (which are potentially regulatory) more readily than non-bound messengers, which is particularly interesting given that glucocorticoids appear to inhibit collagen mRNA translation in lung and dermis by decreasing the amount of polysome-associated (hence cytoskeleton-bound) messenger (Rokowski et al. 1981). In view of this evidence, it seems that mRNA binds to the cytoskeleton immediately on exit from the nucleus, and this cytoskeletal binding is a precondition for both translation and the control of translation. There has been no report so far about the elements of the cytoskeleton that are involved, and there are no indications of specific binding proteins. However, the close association of nuclei with intermediate filaments (Lehto et al. 1978; Staufenbiel and Deppert 1982; Granger and Lazarides 1982), and the reported continuity of the intermediate filament network with the nuclear matrix through the pore-complexes (Capco et al. 1982; Fey et al. 1984) suggests that the intermediate filaments are the likeliest candidates.

The evidence that mRNA transport can only be understood in terms of the solid-state perspective, and that the solution-diffusion perspective is correspondingly misleading, is very varied, as this brief overview illustrates. Collectively, the evidence is weighty. Biochemists and cell biologists find the solid-state perspective difficult to accept on first acquaintance because it seems inconsistent with our intuitions about molecular transport processes. Given modern views about the state of intracellular protein (Fulton 1982) and our knowledge of such solid-state phenomena as axonal transport, we should perhaps be willing to revise such intuitions.

B. An Overview of Nucleocytoplasmic Transport

I. The Nuclear Envelope: Ultrastructure and Biochemistry

My remarks in this section will be brief. They are intended as a summary of our present state of knowledge, and to indicate the points that are most relevant to nucleocytoplasmic transport and the important questions to which we do not, as yet, have satisfactory answers.

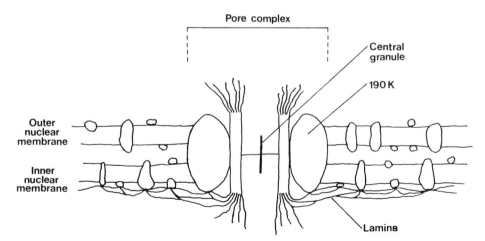

Fig. 2. Scheme of nuclear envelope structure. Excellent micrographs of nuclear envelopes are to be found in the articles by Franke (1973) and Maul (1977) amongst others

The four ultrastructurally distinct components of the nuclear envelope are shown in Fig. 2. The most striking features are the pore-complexes, and these are in many ways the most mysterious. It is generally agreed that these structures are not simple holes, but nevertheless provide aqueous channels through which nucleocytoplasmic exchanges, including macromolecular exchanges, take place. There is electron microscope evidence to support this (Stevens and Swift 1966; Skoglund et al. 1983). The account of RNP particle formation and transport from the Balbiani ring of *Chironomus* by Skoglund et al. (1983) is particularly interesting because it suggests that such transport is associated with changes in RNP tertiary structure (see Sect. D.II). However, Maul (1977) shows that many other pieces of electron microscope evidence for transport through pore-complexes might involve misinterpretation of the data. Also, the export of newly formed virus particles from the nucleus seems to involve the membranes rather than the pore-complexes. It is by such means that viruses can acquire their membranes. Other generally agreed facts about the pore-complexes are their ubiquitous octagonal symmetry, their reproducible size (diameter = approximately 100 nm), and the presence in the pore lumen of a central granule or tubule, occupying about 10% of the total pore-complex diameter (see e.g. Gall 1967). This central granule seems to be an important determinant of the permeability properties of pore-complexes. The mystery attaches first to the detailed architecture of the structures, and second to their biochemistry. More than a score of detailed structural models have been proposed. Four of these, due to Gall (1967), Roberts and Northcote (1970), Franke (1970) and Maul (1982) are shown in Fig. 3, and they illustrate the range and variety of such models. The model given by Maul (1982) is particularly interesting because the "traverse fibrils", roughly orthogonal to the plane of the envelope, might be part of the nuclear-matrix intermediate filament lattice described by Penman's group (see Sect. A.III) and because it incorporates the possibility that pore-complexes are transient structures, formed within the nucleus and passing more or less rapidly through the envelope (see also Kirschner et al. 1977; Engelhardt et al. 1982). This view could be consistent with a model of nucleocytoplasmic transport that involved attachment of the transport substrate to a multimolecular carrier structure, which appears in the envelope as a pore-complex. As for the biochemistry of the pore-complex, there is evidence that in amphibian oocytes it consists of one protein — the major nuclear envelope polypeptide (Krohne et al. 1978a, b, 1982), but other laboratories have not obtained support for this view in oocytes or other systems (Gerace et al. 1978, Aaronson et al. 1982; Maul and Baglia 1983). Indeed, the only unequivocally identified pore-complex component is the 190 kDa glycoprotein which seems to lie on the periphery of the structure, in association with the membranes (Gerace et al. 1982; Filson et al. 1985). This object might correspond to the large particles observed at the peripheries of pore-complexes by Unwin and Milligan (1982), in one of the most detailed ultrastructural accounts of the nuclear envelope to date.

In other respects, our knowledge of nuclear envelope biochemistry is much more substantial, mainly because numerous methods for isolating nuclear envelopes, substantially free of cytoplasmic and intranuclear contaminants, have been developed

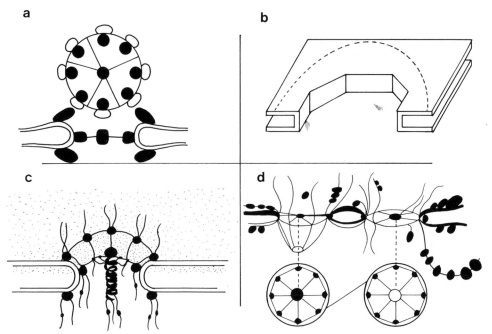

Fig. 3. Models of pore-complex architecture. Four of the many models that have been proposed, viz those by *(a)* Gall (1967), *(b)* Franke (1970), *(c)* Roberts and Northcote (1970) and *(d)* Maul (1982), are shown here. In *(a)* and *(d)*, surface views and sections are shown in juxtaposition. The arrangements of "traverse" fibrils, essentially orthogonal to the membrane plane, are particularly interesting in the light of the "solid-state" perspective on mRNA transport: see especially the Maul (1982) model *(d)*

[see Agutter (1985b) for a discussion of some such methods]. The main protein, carbohydrate and lipid components have been identified and many endogenous enzymes have been described. Most notably, biochemical studies of isolated envelopes, followed by deployment of antibodies against some of their components using tissue sections, have given us a lot of information about the fibrous lamina.

The lamina was first identified as a fibrous network contiguous with the nucleoplasmic face of the inner nuclear membrane, and was isolated in association with the pore-complexes from membrane-depleted nuclear envelopes (Aaronson and Blobel 1975). In many tissues, such as rat liver, it comprises three major polypeptides, which have been given the names lamin A, lamin B and lamin C, in order of decreasing molecular weight (Gerace and Blobel 1980). These proteins lie in the 60–80 kDa range; molecular weight estimates vary quite widely between laboratories. Lamins A and C are have similar sequences and identical isoelectric points, but are apparently encoded on different genes (Laliberté et al. 1984); lamin B is considerably more acidic, has a substantially different amino-acid sequence, is apparently closer to (and perhaps partly buried in) the phospholipid bilayer of the inner membrane, and is much less heavily phosphorylated during interphase (Gerace and Blobel 1980; Shelton et al. 1980; Lebel and Raymond 1984). However, it seems that all three lamins share a large tract of sequence, as indicated by the antigenically common 46 kDa chymotryptic peptide described by Burke et al. (1983). It seems likely that lamins A and C are attached to the membrane through lamin B; moreover, phosphorylation of lamin B is associated with the breakdown of the nuclear envelope during mitosis (Gerace and Blobel 1980). The lamins form homooligomers covalently cross-linked by disulphide bridges, but the weight of evidence is that they are not covalently cross-linked in vivo (Kaufmann et al. 1983). Nevertheless, their propensity for cross-linking may provide clues about the molecular architecture of the lamina in situ. Because of this propensity and because large parts of their surfaces seem to be hydrophobic, they are difficult to solubilize. Cholate (Havre and Evans 1982) or 4M urea (Maul and Baglia 1983) will solubilize the lamins provided that a reducing agent such as 2-mercaptoethanol is present. The former of these treatments apparently yields solubilized dimers with high axial ratios; the state of urea-solubilized lamins is not known. The current state of our knowledge of lamins has been reviewed by Shelton (1985).

The outer nuclear membrane resembles the rough endoplasmic reticulum in composition and structure (Richardson and Maddy 1980). There seems to be some evidence that the inner nuclear membrane is substantially different: its intrinsic polypeptides are unusually immobile, possibly because of close interactions with the lamina, and unlike the outer nuclear membrane it contains phosphatidyl inositol and its derivatives (M. Schindler, pers. commun. to the author; see also Smith and Wells 1984). The immobility of the proteins is consistent with the unusually high-order parameter observed in electron spin resonance studies of nuclear envelopes (Agutter and Suckling 1982a).

The main points made in this brief overview can be summarized as follows:

a) Pore-complexes are mysterious in terms of detailed architecture and biochemistry, but their ubiquitous size and general morphology make it almost certain that they are the sites of macromolecule exchange between nucleus and cytoplasm.
b) Pore complexes might be transient structures, representing movement of multimolecular carrier structures through the proposed nuclear matrix intermediate filament network.
c) Both the pore-complexes and the inner nuclear membrane are attached to the fibrous lamina. One major component of this structure, lamin B, is more closely associated with the inner nuclear membrane than are the other two lamins, and is different from them in its general biochemical properties.
d) Biochemical modification of the lamina (lamin B phosphorylation) is a crucial event in nuclear envelope breakdown during mitosis, indicating that the lamina is essential for the structural stability of the envelope.
e) The inner nuclear membrane has components that might have a role in regulating nucleocytoplasmic transport.
f) The nuclear membranes are probably linked to the pore-complexes by the 190 kDa glycoprotein.

It should be emphasized that the remarks made here neglect the vast majority of our knowledge of nuclear envelope biochemistry. It will become appropriate to mention one or two more elements of this knowledge later in this article. At present, I have deliberately restricted the discussion to those aspects of the topic that are most obviously pertinent to nucleocytoplasmic transport.

II. In Situ Studies of Transport

1. Nuclear Envelope Permeability

From the remarks made about pore-complexes in the previous subsection, it will be clear that the nuclear envelope is likely to behave as a molecular sieve. Objects will diffuse through it if they have hydrated diameters less than the patent diameters of the pores. This prediction has been thoroughly corroborated by a range of permeability studies using inert permeants and exogenous labelled proteins of known size, and such studies have also made it possible to calculate the patent pore diameter, at least approximately. The inert permeants, such as colloidal gold, ferritin and tritiated dextrans, are microinjected into the cells under study. For obvious technical reasons, giant cells — amphibian oocytes and *Amoeba proteus* — have been used in most in situ experiments. More recently, using more sophisticated techniques, smaller cells, including hepatocytes and HeLa cells, have been investigated.

Feldherr (1972) reviewed the studies that he pioneered during the 1960's on the movement from cytoplasm to nucleus of microinjected colloidal gold and ferritin. In these studies, the location of the permeant was determined by electron microscopy of thin sections at various times after microinjection. These experiments established the pore-complexes as the sites of nucleocytoplasmic exchange of colloidal particles, and indicated an inverse relationship between permeation rate and particle size. Quantitative studies of this relationship were conducted by Paine et al. (1975). Their experiments involved cryofixation and autoradiography of *Xenopus* oocytes at various times after microinjection of tritiated dextran spheres of known diameter, and the results indicated a patent pore diameter of about 9 nm. This is about the size of the central granule or tubule of the pore-complex. This correspondence is interesting, but Paine and his colleagues introduced a number of assumptions into their calculations, duly emphasizing that these were open to doubt. However, in separate studies, various fluorescently labelled proteins were also microinjected into oocytes and insect salivary gland cells, and the rates of movement of fluorescence into the nuclei were again found to be consistent with a patent pore diameter of 9 nm (Paine and Feldherr 1972; Paine 1975).

More recently, fluorescence microphotolysis (photobleaching) has been used to measure nucleocytoplasmic transport rates and to estimate patent pore radii in smaller cells as well as oocytes. After the microinjected label has equilibrated between the cell compartments, the cell, located on a glass slide on a microscope stage, is irradiated by a laser beam for a few milliseconds so that the fluorescence in the nucleus is partially depleted. The recovery of fluorescence in the nucleus then provides a sensitive indicator of flux rates across the envelope. The technique obviously requires fluorescent molecules that can be irreversibly bleached. Given these, it is invaluable. It has a very high spatial resolution (μm) and time resolution (ms), requires only small quantities of materials and small numbers of cells, and is rapidly performed. The results (Peters 1983a, b; Lang and Peters 1984) have shown that in hepatocytes and HeLa cells the patent pore diameters are slightly but significantly greater than those in oocytes (10–11 nm), and that some pore-complexes are damaged, losing their central granules and becoming more or less empty holes, when nuclei are isolated or resealed nuclear envelope vesicles are prepared. These findings have very important implications for in vitro studies of transport (see next subsection). Even when only a small percentage of the pore-complexes are damaged in this way, the permeability properties of the envelope are markedly perturbed. Indeed, as Lang and Peters (1984) showed, these permeability measurements provide a means of assessing envelope integrity that is unprecedentedly sensitive.

The predictions about nucleocytoplasmic transport that can be made from such permeability studies are as follows. First, the half-times of diffusion equilibrium of spherical molecules across the nuclear envelope will vary inversely with the size of the molecule. In effect, globular proteins with molecular weights up to about 5 kDa will diffuse at roughly the same rates as in free solution, while proteins of around 60 kDa will be more or less non-permeating. RNP complexes, including messengers, are too big to diffuse through the molecular sieve of the envelope. These predictions are realized in vivo to the extent that many low molecular weight proteins have nucleocytoplasmic concentration ratios in the order of unity. However, some very low molecular weight species, such as histones and high mobility group proteins, are highly karyophilic (nucleocytoplasmic ratios in excess of 500:1); some very high molecular weight polypeptides, such as the core components of RNA polymerases, also enter and are concentrated in the nucleus; some proteins, such as tubulin, do not enter the interphase nucleus at all; and mRNP (and ribosomal subunits) do, as a matter of fact, pass into the cytoplasm. Thus, although the physical permeability properties of the nuclear pore-complexes are important determinants of protein flux and distribution, they do not explain everything about nucleocytoplasmic transport. There must, in the large transportable and in the small karyophilic and cytoplasm-restricted molecules, be signals for active uptake through the pores or for accumulation in either the cytoplasm or the nucleus.

2. Protein, tRNA and snRNP Transport

Notwithstanding the importance of pore-complex permeability in determining protein flux rates, the fact remains that nucleocytoplasmic concentration ratios of proteins vary from the immeasurably high to the immeasurably low, and the correlation between this ratio and protein molecular weight is far from perfect. Our understanding of the reasons for this has advanced considerably in the last 5 years, and in the next few paragraphs I shall survey the current progress of this understanding, together with parallel developments in the analysis of tRNA and snRNP transport.

Relevant background studies began in the late 1960's and by 1980 a considerable body of data had accumulated. However, although these data led to some interesting generalizations, they lacked a mechanistic explanation (Paine and Horowitz 1980). The main findings were as follows. First, when proteins from cells of a particular type (usually *Xenopus* oocytes) were microinjected into intact living cells of the same type, they rapidly and efficiently achieved their normal in vivo distributions between nucleus and cytoplasm. Thus, Gurdon (1970) showed that radioiodinated histones rapidly concentrated in the nucleus after microinjection; and the same was found for HMG1 (Rechsteiner and Kuehl 1979) and other non-histone chromosomal proteins (Yamaizumi et al. 1978). Second, when foreign nuclei were implanted into cells, some proteins shuttled rapidly between the nuclei, and these were the same proteins that accumulated rapidly in the nuclei in microinjection experiments (Legname and Goldstein 1972; Jelinek and Goldstein 1973). Third, as expected from the permeability studies, oocyte nuclei allowed free entry of small proteins and excluded large ones (Bonner 1975a), but there was a selectivity other than that based on size. Not only did microinjected karyophilic proteins accumulate in the nucleus, but microinjected cytoplasmic proteins remained outside the nucleus (Bonner 1975b). Fourth, the differences and similarities in nuclear and cytoplasmic protein compositions, well described by Feldherr (1975), were investigated further by De Robertis et al. (1978) using the then-recent technique of 2d gel electrophoresis. Their results emphasized the rapidity with which karyophilic proteins accumulate in the nucleus, and led them to infer that karyophilic proteins must contain a signal responsible for such accumulation. Fifth, in view of the fact that karyophilic (and cytoskeletal) proteins do not, like most proteins, redistribute when the nuclear envelope is torn (Feldherr and Ogburn 1980), it became accepted that the signal acts by binding the protein to some large immobile intranuclear structure, not by facilitating passage through the pore-complex. Sixth, mature proteins, not precursors with a "signal" sequence that is later removed, enter the nucleus (Wu and Warner 1971; Dabauville and Franke 1982). Seventh, nucleocytoplasmic separations of most proteins cease to exist during mitotic prophase and are re-established at telophase (Beck 1962). It now seems that most or all of these generalizations apply to cells other than oocytes and amoebae (e.g. Stacey and Allfrey 1984).

Since 1980, there has been progress in characterizing the karyophilic signals. Nucleoplasmin, the most abundant karyophilic protein, is a pentamer in which each 40 kDa monomer has a "head" and a "tail" region. It appears to have a role in nucleosome assembly, presenting histones to the DNA (Laskey and Earnshaw 1980; Earnshaw et al. 1980). The "tail" regions contain the karyophilic signal (Dingwall et al. 1982). If all the tails are removed from the pentamer, leaving a core of five "head" regions, the core is excluded from the nucleus. Isolated tails accumulate in the nucleus. If one to four of the tails are removed, nucleoplasmin still accumulates in the nucleus, but at a rate proportional to the number of tails remaining. However, if the tail-less core is injected directly into the nucleus, it remains there. This last result may cast doubt on the belief that signals are responsible for binding and retention within the nucleus, not for entry. The findings by Kalderon et al. (1984) concerning the large-T antigen of SV40 might cause further doubt. Nuclear localization of large-T depends on amino-acid residues 127–131 — a strikingly basic sequence (lys-lys-lys-arg-lys), alterations of which, especially those involving lys-128, cause retention of the antigen in the cytoplasm. However, alterations of other, quite remote, parts of the molecule, while not preventing nuclear entry, do alter its final nucleocytoplasmic distribution. More explicitly, Davey et al. (1985) found that a non-terminal 19-residue sequence is responsible for the uptake of influenza virus nucleoprotein in *Xenopus* oocytes, but accumulation still occurred when this sequence was absent. A possible implication of this work is that two sorts of signals exist: pore-complex premeation signals, and intranuclear accumulation signals. Presumably the former, but not the latter, would be redundant in low molecular weight proteins that could enter the nucleus rapidly by diffusion.

It now seems clear that some of the differences between actual nucleocytoplasmic distributions and accumulation rates of proteins, and those predicted from the permeability properties of envelopes, can be explained by such signals. The field is still immature, but complete characterization of some of these signals can be expected soon. On the basis of the large-T antigen evidence and the comments by Paine and Horowitz (1982), charge — presumably positive — might prove to be a common feature of accumulation signals. However, not all the anomalous distributions and fluxes can be explained in terms of intramolecular signals of this kind. Protein binding in the cytoplasm is also important. Just how important was revealed only when the "reference phase" technique was introduced by Paine, Horowitz and their colleagues. Briefly, this technique involves microinjection of gelatin (approximately the same volume as the nucleus) into an oocyte, allowing time for diffusible proteins to enter it, and then isolating the nucleus, the cytoplasm and the reference phase (i.e. the gelatin) by cryodissection (i.e. rapid freezing in liquid nitrogen followed by microdissection of the frozen cell contents). Cryodissection is essential to prevent interference from diffusion effects during manipulation of the cell. The results have not only confirmed the in vivo differences between nuclear and cytoplasmic protein contents, but have revealed that many cytoplasmic proteins have only limited diffusibility; some, indeed, are completely immobile. Obviously, only diffusible proteins can enter the reference phase (Austerberry and Paine 1982; Paine 1982; Paine et al. 1983). As for the less diffusible and the immobile proteins, Paine (1985) attributes their lack of free mobility to their association with the cytoskeleton. This is certainly the correct explanation in most cases, but there are other possibilities. For instance, the 40 kDa 5S-RNA binding protein can enter the nucleus readily when it is free, but not when it is RNA-bound (Engelke et al. 1980). Here, the restricting factor is not the cytoskeleton, but a particular RNA species.

RNA-protein interactions play a different kind of role in snRNP accumulation in nuclei. When U1, U2, U4, U5 or U6 snRNA are injected into oocytes they become highly concentrated in the nucleus, unlike tRNA and 7S RNA, which remain in the cytoplasm (De Robertis et al. 1982). Specific association of these snRNA's with proteins is crucial for the rapid transport and accumulation. It appears that the associated proteins are stockpiled in the cytoplasm during oogenesis and early development (Zeller et al. 1983), and there is some evidence that they bind by recognizing certain specific secondary-structure domains in the snRNA itself (Forbes et al. 1983). When endogenous snRNA synthesis begins at the mid-blastula stage, the naked RNA apparently leaves the nucleus, binds proteins from the stockpile and very rapidly re-accumulates in the nucleus (see e.g. Mattaj and De Robertis 1985). This movement is not sensitive to agents that disrupt the cytoskeleton (Zieve 1984). The close coupling between nucleocytoplasmic snRNA migration and associated protein migration might exemplify a general phenomenon, applying to RNA classes other than small-nuclear (Mills and Bell 1982). If this is so, then there are clear possibilities for controlling cellular activities at the level of RNA transport.

Export of tRNA to the cytoplasm is clearly controllable, because in rabbit reticulocytes the cytoplasmic tRNA content is specialized for the amino-acid composition of globin (Smith and McNamara 1971). This control might be exerted at the level of precursor processing and detachment from the intranuclear binding proteins (cf. Rinke and Steitz 1982), but it might alternatively act through recognition of specific intramolecular signals by the pore-complex. By microinjecting mutants of human initiator tRNA-met, having one or more C-T transitions, into *Xenopus* oocyte nuclei, Tobian et al. (1984) showed that tRNA processing and nucleocytoplasmic transport are affected independently. One nucleotide, G-57, was particularly crucial for transport: replacement of it by C resulted in a 50-fold decrease in transport rate (Zasloff 1983). Generally, the first 62 nucleotides in the molecule seem to contain the sequences regulating transcription, processing and transport (Adeniyi-Jones et al. 1984). Nucleocytoplasmic transport of tRNA involves a specific facilitated diffusion mechanism, showing saturability, sequence-specificity, and competition between tRNAs (Zasloff 1983). It is notable that the region of the molecule that is important for transport is particularly highly conserved.

So far, I have made the following main points about our current state of knowledge in this field:

a) Many aspects of nucleocytoplasmic protein distribution and flux can be accounted for by the specific permeability properties of the pore-complex.

b) Aspects that cannot be so explained have been accounted for by:

1. protein binding to the cytoskeleton, to cytoplasm-restricted RNA's, etc;
2. intramolecular signals that bind the protein to immobile intranuclear sites;
3. possibly signals that are recognized by the pore-complex and participate in facilitated diffusion;
4. association with RNAs that cross the nuclear envelope.

c) Some RNA-protein complexes can cross the nuclear envelope and accumulate in the nucleus, although the proteins alone cannot, suggesting that signals for both binding and diffusion can be more than short amino-acid sequences. Secondary (and quaternary) structure characteristics might be important.

d) RNA transport across the nuclear envelope also involves signals, probably of both intranuclear-association and pore-recognition types. So far, there is no direct evidence that association with proteins is necessary for RNA movement across the envelope, though it certainly seems that association with RNA is necessary for the migration of certain proteins.

e) Just as some proteins seem to be immobilized in one compartment or the other by RNA association, so, possibly, RNAs might be immobilized by association with particular proteins.

f) Nucleocytoplasmic transport processes might be controllable, and their regulation might be important in controlling cellular function in general.

I shall take up these points in Section B.II.3, below. To end the present discussion, I wish to develop point (f) by drawing attention to the functional significance of some nucleocytoplasmic protein exchanges. These are, potentially, relevant to a discussion of the biological significance of mRNA transport (Sect. E).

First, it is perfectly clear that exchange of proteins between mobile and immobile phases, notably the core proteins of cytoskeletal elements, is itself functionally significant in development (Olmsted 1981; Korn 1982). Since mobile proteins can potentially enter the nucleus and immobile ones cannot, this developmental significance is related (at least coincidentally!) to the regulation of nucleocytoplasmic exchange. Secondly, and more significantly, the reprogramming of transplanted nuclei involves uptake into the foreign nucleus of proteins from the host cytoplasm (Appels et al. 1975; Hoffner and Di Bernardino 1977). Somatic nuclei injected into oocytes tend to express the genes that the uninjected oocyte expresses (Gurdon and Melton 1981). Gurdon and his colleagues have also used the developmental switching of amphibian 5S RNA genes as a model to examine this phenomenon. Oocyte 5S genes are switched off during development. Normal oocyte extracts stimulate the re-expression of these genes in erythrocytes from the adult animal, and this effect is sensitive to heat and to proteinases. Moreover, some oocytes are unable to inactivate 5S genes, and their cytoplasms fail to stimulate the expression of oocyte-type 5S genes in the erythrocyte (Korn et al. 1982). Thirdly, various proteins, detectable by specific antibodies, pass from the oocyte nucleus to the cytoplasm during meiosis. These proteins re-enter the nuclei individually, at different stages of development, presumably controlling changes in gene expression that are proper to the various developmental stages (Dreyer et al. 1983). On the basis of these findings, it seems clear that nucleocytoplasmic protein movements play a very important part in the control of gene expression and of cellular differentiation. The fact that we are as yet unable to explain the mechanisms underlying the developmental stage-specific importing of proteins, described by Dreyer et al. (1983), highlights our ignorance of many crucial facts about nucleocytoplasmic transport, which persist despite the apparently rapid progress in understanding over the last 5 years, and the elegance of the techniques that are now being deployed.

3. Implications for mRNA Transport

Although in situ studies have not directly thrown any light on the issue of mRNA transport, the findings pertinent to proteins, tRNA and snRNPs have implications that could usefully be borne in mind when the discussion in Section D is considered. If the likelihood that mRNA transport is a solid-state process, fundamentally different from the transport processes considered so far, is ignored for the moment, the following inferences can be derived from the preceding discussion:

a) Messenger RNA or RNP is too large to leave the nucleus by passive diffusion through the pore-complexes. It almost certainly exits via the pore-complexes, but the physical (permeability) properties of these structures are not directly relevant.
b) It would not be surprising if messengers were equipped with intramolecular signals which (1) ensured binding to sites in the nucleus and the cytoplasm, (2) bound to the pore-complex, where they had a role in facilitated or active transport.
c) Given that the protein components of HnRNP and mRNP are different, it is possible that these components are crucial in retaining messenger precursors in the nucleus and mature messengers in the cytoplasm.
d) Introns, presumably by linkage to specific proteins, may act as nuclear binding or accumulation signals.
e) It is reasonable to expect that some stage in mRNA transport, perhaps translocation through the pore-complex, is controllable.

If we now recall the likely solid-state nature of mRNA transport, it will readily be seen that these inferences do not become invalid. Rather, it becomes appropriate to state some of them [notably (a) and (b)] more strongly. This position in turn has important implications. It seems likely that introns in messenger precursors, common sequences (or secondary-structure regions) that are potential "signals" in mature messengers, and association of HnRNA and mRNA with specific proteins, are all likely to be relevant to mRNA transport. Therefore, in order to understand mRNA transport, it is first necessary to understand RNA metabolism and ribonucleoprotein organization.

Another kind of inference that can be drawn from the preceding discussion concerns the intractability of transport mechanisms. Despite the partial characterization of signals, despite the measurements of flux rates, nucleocytoplasmic distribution ratios, and patent pore diameters, and despite the wealth of elegant techniques that have been devised, the following important factors have so far limited progress:

1. The pore-complex is not biochemically characterized.

2. Neither are other potentially important structures, such as the putative nuclear matrix.
3. Because living cells have large ATP pools and at least some capacity for anaerobic metabolism, the energy-dependence of transport processes cannot be ascertained by any available techniques. This is true not least of amphibian oocytes.
4. Intracellular homeostasis precludes the possibility of constraining transport systems in the living cell in the kinds of ways that would usefully elucidate mechanisms (e. g. by altering the ionic conditions or introducing toxic inhibitors during experiments with long time-courses).
5. The sheer complexity of intracellular composition, organization and physical state put limits on the interpretation of observations at the biochemical level.

It is obvious that although points (1) and (2) are general barriers to progress, points (3), (4) and (5) point specifically to the inherent limitations of in situ systems. They clearly indicate the need for satisfactory in vitro systems. The limitations, and the successes, of such systems are indicated in the following subsection.

III. In Vitro Studies of RNA Efflux

There are two kinds of in vitro systems for studying nucleocytoplasmic transport: those that involve isolating nuclei, incubating them in a suitable medium, and examining the post-nuclear supernatant after incubation; and those that involve preparing resealed nuclear envelope vesicles, in which transportable or untransportable materials are trapped during resealing, and again examining the movement of these materials into the supernatant during incubation. Both approaches are inherently problematic, for the following reasons.

In Sections B.II.1 and B.II.2 of this section, I mentioned two findings that throw very serious doubt on the possibility of valid, interpretable in vitro studies. First, the finding by Lang and Peters (1984) that at least some pore-complexes lose their central granules when nuclei are isolated, and still more when envelopes are isolated, makes it clear that envelopes in in vitro systems always have non-physiological permeability properties. Second, Paine et al. (1983) showed that when oocyte nuclei are isolated by the usual aqueous procedures, 95% of their protein is lost with a half-time of about 4 min. Even allowing for effects of proteolytic degradation (which can largely be prevented, in practice, with serine proteinase inhibitors) in these studies, this leakage during cell-fractionation and nuclear isolation is obviously serious. The authors make the point that with smaller nuclei, such as hepatocyte nuclei, the loss of contents will be still more rapid. Apart from these two very serious sources of doubt, other general criticisms can be added. For instance, the close association of intermediate filaments with the nuclear surface might be very significant for transport processes, and this association is lost or seriously perturbed when nuclei are isolated. Also, nuclear envelopes are sometimes grossly torn during nuclear isolation. Thus, the results of in vitro "transport" studies could be attributed to leakage from a few badly-damaged nuclei, rather than to physiological efflux from all of them.

Against these objections, one has to set the urgent need for adequate in vitro systems that was indicated at the end of the previous subsection. The effect of this conflict over the last decade has been as follows. Proponents of in vitro systems for studying mRNA efflux have spent a good deal of time in trying to justify their use of such systems and to show that their results are physiologically significant. Some of these "justifications" have been inadequate and have merely fuelled the controversy. However, some in vitro systems are now well-characterized, and provided that the results are interpreted within the established limitations of the methods, they need no further justification (see below). At the same time, workers who are not proponents of such systems have been aware of the serious reasons for doubting the value of in vitro methods for any study of nuclear transport processes. These workers have often taken the view that in vitro systems simply cannot be justified. This position, of course, terminates any rational discussion of the matter, but fails to end the controversy.

I discussed the problems of in vitro systems at length in an earlier review (Agutter 1984), and there would be no point in repeating the details here. Instead, I want to recall the main points at issue, to show how recent characterization studies have validated such methods, and to indicate the extent and the limits of this validation.

1. Isolated Nuclei

The paper that is crucial in validating studies of mRNA efflux from isolated nuclei is that of Jacobs and Birnie (1982). Using cDNA probes against cytoplasmic messengers and against nucleus-restricted sequences, these authors showed that (a) the messenger population in the post-nuclear supernatant was distributed into the three abundance classes characteristic of cytoplasmic messengers, (b) significant contamination with immature mRNA precursors amounted to no more than 10% of the eluted RNA and was restricted to the lowest abundance-class species. These findings applied to hepatocyte and hepatoma nuclei incubated in the medium devised by Yu et al. (1972). Otegui and Patterson (1981) showed when that isolated myeloma nuclei were incubated in a very different medium, only mature immunoglobulin messengers, not precursors, were eluted into the supernatant. Once again, messengers in the highest abundance class appear exclusively in their mature form in such experiments.

Some supplementary evidence can be adduced in support of these centrally important findings. Messengers in the post-incubation supernatants can be translated with normal kinetics to give the proteins made in vivo. This was definitively shown by Palayoor et al. (1981), but was already strongly suggested by the pioneering studies of Ishikawa et al. (1970a, b 1972). Intranuclear proteins, including the HnRNA core proteins, do not appear in the supernatants (Agutter 1982).

Among the criticisms that have been made of in vitro studies have been the following. (a) The supernatant RNA might have desorbed from the nuclear surface, rather than coming from the intranuclear compartment. However, both Jacobs and Birnie (1982) and Agutter (1983) found that contamination from surface-desorbed RNA amounted to no more than 10–15% of the supernatant material. (b) The supernatant RNA might result from degradation of intranuclear RNA rather than from physiological processing. Degradation is indeed a problem, but can be prevented by including carrier RNA (Yu et al. 1972), cytoplasmic ribonuclease inhibitor (Yu et al. 1972; Clawson et al. 1978) and serine proteinase inhibitors (Agutter 1983) in the medium. (c) The RNA might leave the nucleus non-physiologically, through tears in the envelope, rather than through the pore-complexes. However, apart from the fact that Feldherr's (1980) results make this seem improbable, it has been found that high concentrations of colchicine, which constrict the pore-complexes while disordering the nuclear membrane phospholipid bilayers, prevent RNA efflux in vitro (Agutter and Suckling 1982b; Schumm and Webb 1982). More definitive evidence on this point, if it were necessary, would require specific antibodies against pore-complex components. In the present state of characterization of the pore-complex, however, these are not available.

It can be added that when mRNA effluxes from two or more types of nuclei are compared in vitro, the differences in the supernatant contents reflect the differences in cytoplasmic mRNA compositions found in vivo. These effects are seen, for instance, on carcinogen feeding of rats (Schumm et al. 1973; Smuckler and Koplitz 1974, 1976; Lemaire et al. 1981), adenovirus infection (Raskas and Bhaduri 1973) and tryptophan feeding (Murty et al. 1980). There is also incidental support for in vitro systems from competition-hybridization studies, morphology of the nuclei before and after incubation, and other indicators (see discussion in Agutter 1984).

All these results make it clear that, granted certain limitations, in vitro systems for the study of mRNA efflux are valid and interpretable, representing physiological phenomena. The "certain limitations", however, are important:

a) As yet, isolated nuclei cannot be used for studying the transport of low abundance class messengers, at least not without risk of serious interference.

b) The source (cell-type) of the nuclei, and the composition of the medium, are both important. Complete characterization of mRNA efflux needs to be carried out for each new source of nuclei and each new medium. The encouraging findings to date should not be over-generalized.

c) Insofar as the cytoskeleton plays a part in mRNA transport, in vitro methods, in which the cytoskeleton is absent, do not reflect transport. In terms of the nomenclature introduced in Section A.II, a satisfactory in vitro system examines release + translocation, not cytoskeletal binding. The fact that physiological results are obtained indicates that cytoskeletal binding plays no significant part in selecting the messengers for export from the nuclei, but it could be significant in other respects, e.g. regulation of transport.

Finally, it should be re-emphasized that such in vitro systems, though it is now clear that they can be used within these limits as models for studying mRNA transport, are of no value for studying (e. g.) protein transport. This underlines the differences between protein and mRNA transport processes, and draws attention once again to the relative applicabilities of the solution-diffusion and solid-state perspectives.

2. Resealed Vesicles

Studies with isolated nuclei, it seems, throw useful light on the mechanisms of release + translocation. It is also useful to have a system in which translocation alone can be studied, i. e. in which nuclear-envelope events can be observed without interference from intranuclear events. The most significant progress in this field to date has come from the work of Fasold and his colleagues (Kondor-Koch et al. 1982; Riedel et al. 1987).

In these studies, hepatocyte nuclear envelopes have been resealed in the presence of 3 mM calcium, forming vesicles with about the same diameters as the original nuclei. During the resealing, solutes can be trapped in the vesicles. During subsequent incubations, the rates of egress of trapped materials, and the rates of entry of macromolecules added to the supernatant, can be measured. The effects of ATP, temperature, ionic conditions and various inhibitors can easily be examined. All these circumstances provide important advantages over in situ approaches.

Despite the leakiness of these vesicles predicted by the studies of Lang and Peters (1984), some highly interesting and significant results have been obtained. Translocation through intact pore complexes seems to outstrip diffusion-leakage so greatly in terms of both rate and extent that physiologically interesting effects are easily detected against the background of leakage. Indeed, the results suggest that only 10–15 % of the vesicles are leaky (Riedel et al. 1987). The most intriguing results are as follows. High molecular weight, non-karyophilic proteins such as IgG do not leave the vesicles when they are trapped. The vesicles rapidly accumulate histones to a concentration 25 times that in the supernatant, and they achieve this in the absence of nucleoplasmin. They do not, however, concentrate other low molecular weight basic proteins such as cytochrome c, which are not associated with nuclei in vivo. Messenger RNA leaves the vesicles by a mechanism dependent on ATP hydrolysis, and a poly(A) tail greatly accelerates the efflux (see Sect. D for a fuller discussion).

One question that can be raised about these findings concerns the persistence within the vesicles of relevant intranuclear binding sites. For instance, if all the histone binding sites have indeed been extracted during vesicle preparation, then the accumulation of histones must presumably be attributable to pore-complex events. If so, then some rethinking of our present views about the "signals" involved in protein transport will be necessary. However, this point remains to be clarified. Such doubts are inevitable when methods are new, and in this case will no doubt be resolved in the near future. Despite them, it already seems clear that this technique is an invaluable addition to our armoury of methods for studying nucleocytoplasmic transport, and we may reasonably expect it to make important contributions to progress in the field over the next few years.

C. RNA Metabolism and Ribonucleoprotein Structure

In the last section, I drew attention to the particular relevance of mRNA metabolism and of mRNP and HnRNP structure to an understanding of mRNA transport. In an excellent review published 5 years ago, Martin et al. (1980) made the point that progress with our analyses of RNP structure has always depended on progress in our understanding of RNA metabolism. To some extent this remains true, and I shall therefore consider HnRNA and mRNA metabolism in this section before I discuss RNP structures. (Incidentally, comparison of that 1980 review with our current knowledge gives a good indication of the rate and extent of recent progress). Following the survey of HnRNP and mRNP structures, I shall review the nuclear matrix controversy. This can be seen as an extension to the discussion of HnRNP organization and, at the same time, as an examination of what the solid-state perspective implies.

I. Post-Transcriptional Processing

In this subsection, I shall consider developments in our understanding of capping, adenylation and splicing of HnRNA, with emphasis on aspects of these processes that seem most relevant to transport. I shall add a section on the further metabolism and degradation of cytoplasmic mRNA. This is not usually considered an aspect of post-transcriptional processing, but is included here because it might be relevant to the cytoskeletal binding stage of mRNA transport, and because it is certainly important in the control of metabolism (see also Sect. E). This general overview of processing is summarized in Fig. 4.

1. Capping

Capping is the methylation of the 5'-terminal base (G) and ribose, and occurs very rapidly on the nascent RNA transcript. Three biological functions have been ascribed to it. (a) It seems to have a role in initiating translation (Shatkin 1976; Filipowicz 1978; Bannerjee 1980). These authors argue not that it is an absolute requirement for forming functional initiation complexes, but that it accelerates initiation five- or tenfold. (b) It seems to protect HnRNA and mRNA against degradation by 5'-nucleases (Furuichi et al. 1977; Shimotahno et al. 1977). (c) There is growing evidence that it is necessary for efficient splicing (Konarska et al. 1984).

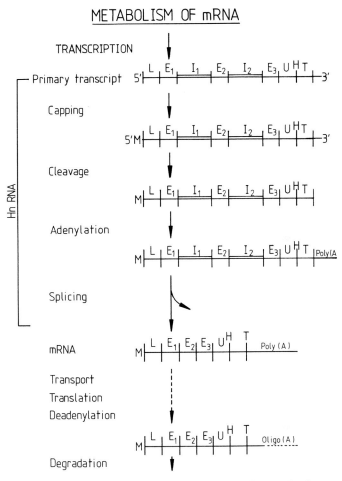

Fig. 4. Summary of the events in HnRNA processing and mRNA catabolism. *L* leader sequence; *E* exons; *I* introns; *U* untranslated region; *H* the hexanucleotide AAUAAA; *T* 3'-terminus that is removed during cleavage; *M* methyl cap

Particularly interesting in terms of the present article is the existence of polypeptides that bind specifically to methyl-capped 5'-termini. Sonenberg and Shatkin (1977) identified such a protein by covalent cross-linking of RNA to closely associated polypeptides, using high-intensity UV irradiation. (This photoaffinity labelling technique is very important in recent studies of RNA-protein interactions, and it will be mentioned several times in this chapter). Sonenberg's group reported the purification of this protein 3 years later (Traschel et al. 1980). Hellmann et al. (1982) identified a 24-kDa polypeptide that reversed the inhibition of cell-free translation by cap analogues, and Rhoades and his colleagues subsequently developed an efficient new purification procedure (Webb et al. 1984). There is no doubt that this 24-kDa polypeptide is an important cytoplasmic cap-binding protein, identical to that reported by

Sonenberg's group, and probably necessary for translation to be initiated efficiently (Shatkin 1985). Studies are now in progress to elucidate the detailed biochemistry of its interaction with the cap. However, it might not be the only cap-binding protein. Chakraborty et al. (1982) found that antibodies against it cross-reacted with several other mRNA-associated polypeptides that were differently distributed amongst the free and polysome-bound messenger populations in embryonic muscle cells. In the same year, Patzelt et al. (1982) identified five distinct nuclear, as opposed to cytoplasmic, cap-binding proteins in HeLa cells. They, again, used the photoaffinity labelling technique. The five proteins had molecular weights of 120, 89, 80, 37 and 24 kDa; the last is, presumably, identical with the major cytoplasmic cap-binding protein.

The possible relevance of these studies to mRNA transport is as follows. Since Zumbe et al. (1982) found the major cap-binding protein of baby hamster kidney cells to be associated with the intermediate filaments, the existence of the cap and of its specifically associated protein(s) might be crucial for cytoskeletal binding. Since caps are important for efficient translation, and since cytoskeleton-association seems to be a prerequisite for translation (see Sect. A.III), this relationship might be biologically important (cf. Agutter and Thomson 1984). By analogy, one might speculate that intranuclear cap-binding proteins are involved in binding HnRNA to the nuclear matrix. (Actually, Zumbe et al. (1982) reported a molecular weight of 50 kDa for their cap-binding protein, but apparently its tryptic peptide map revealed a close structural similarity to the better-known 24 kDa species).

It should be added that non-terminal methylations of HnRNA might also be important at some stage of mRNA transport. In late SV40 mRNA, methylations of internal adenine residues seem to have a role in modulating the processing-dependent nucleocytoplasmic transport (Finkel and Groner 1983). At this stage, however, the relevant mechanisms remain obscure.

2. Adenylation

Close to the 3'-end of primary RNA transcripts is a highly conserved hexanucleotide, AAUAAA, which constitutes the site of adenylation (Proudfoot and Brownlee 1976). Transcription actually proceeds downstream of this hexanucleotide by some 35 residues (Ford and Hsu 1978; Nevins and Darnell 1978; Fraser et al. 1979; Hofer and Darnell 1981). This additional 3' sequence is necessary for accurate cleavage at the hexanucleotide, and hence for subsequent adenylation (Hofer et al. 1982; McDevitt et al. 1984). When transcription is inefficiently terminated, adenylation is also inefficient (Acheson 1984). Adenylation occurs quite rapidly after transcription, and normally precedes splicing (Nevins and Darnell 1978; but see Darnell 1982).

The best-established biological role of adenylation, and according to some workers the only role, is in the maintenance of messenger stability. The evidence includes the following points. Messengers in situ that have short poly(A) tails, or no poly(A) tails, tend to have shorter biological half-lives than those with normal tails — indeed, half-life seems to correlate positively with tail length. During normal messenger degradation, shortening of the poly(A) tail is commonly observed as an early event. When messengers are microinjected into cells, their subsequent life-spans are normal if their poly(A) tails are intact, but short if they have been artificially truncated. These various lines of evidence (see e.g. Marbaix et al. 1975; Nudel et al. 1976; Huez et al. 1981) are discussed in reviews by Littauer and Soreq (1982) and Mueller et al. (1985). The issue will be discussed further in Section C.I.4, below. Another role that has been suggested for poly(A) tails is in splicing (Bina et al. 1980) in which it could participate by activating endoribonuclease VII (Bachmann et al. 1984a). In this context, it is interesting that antibodies against U1-snRNP inhibit adenylation as well as splicing (Moore and Sharp 1985), but this suggested role requires further critical study.

The most controversial suggestion about a role for adenylation, and the one most relevant to the present review, concerns transport. Shortly after poly(A) tails were discovered, a role in nucleocytoplasmic messenger transport was suggested (Adesnik et al. 1972; Jelinek et al. 1973). Some fairly strong evidence in support of this suggestion has been described at various times since these papers appeared. For instance, Villarreal and Whyte (1983) studied a splice-junction deletion mutant of SV40 and found that it produced a mixture of processed RNAs. The only obvious common structural feature of these was a complete lack of poly(A), and none of them was transported to the host cytoplasm. Studies on the mRNA translocation system in isolated nuclear envelopes have indicated that a RNA binding site is important, interacting functionally with the energy-transducing system, and this site has at least some specificity for poly(A). This evidence will be discussed more fully in section D. Moreover, Riedel et al. (1987) showed that poly(A)+RNA was much more rapidly and efficiently extruded by resealed nuclear envelope vesicles than poly(A)-RNA.

However, there are reasons for doubting that poly(A) has anything to do with messenger transport. First, it is clearly not a sufficient condition for transport in itself, since many poly(A)+RNA species remain restricted to the nucleus in vivo (see e.g. Herman et al. 1976; Kleiman et al. 1977; Lasky et al. 1978; Salditt-Georgieff and Darnell 1982). This is not a very serious reservation: it is inherently unlikely that mRNA transport could be controlled by satisfaction of a single condition, such as possession of a poly(A) tail. The process is surely more complex. Second, however, there is evidence that it is not even a *necessary* condition for transport. Two lines of evidence are important here. (a) Histone messengers, except in some specialized cell types, do not have poly(A) tails (see e.g. Ballantine and Woodland 1985). They have correspondingly short biological half-lives; but there is no doubt that they are transported efficiently to the cytoplasm. (b) When adenylation is inhibited in cultured cells by 3'-deoxyadenosine (cordycepin), adenovirus mRNA still enters the host cytoplasm and the polysomes, though of course its biological half-life is shorter (Zeevi et al. 1982). Counter-arguments can be offered to these lines of evidence. For instance, the nuclear envelope binding site has a general affinity for oligopurinosines, including poly(A) (Agutter et al. 1977), and histone messengers obviously contain extensive oligopurinosine sequences. Cordycepin has effects on mRNA efflux in vitro that cannot be ascribed to inhibition of adenylation (Agutter and McCaldin 1979; Kletzein 1980). In any case, it is not clear from the findings of Zeevi et al. (1982) that the poly(A) tail was completely absent in the cordycepin-treated cells; it is only clear that it was abnormally short, and it is known that oligo(A) tracts as short as 15 residues can interact efficiently with the nuclear envelope system (Agutter et al. 1977; Bernd et al. 1982 a).

It remains feasible that adenylation, like capping, is a multifunctional process. Despite the apparently weighty arguments against the need for poly(A) tails in nucleocytoplasmic mRNA transport, it seems to me that the weight of evidence currently favours such a role. However, the requirement is not absolute. It is reasonable to suppose that some part of the messenger molecule — a permeation "signal" — has to interact with the nuclear envelope binding site and that in most, but not all, messengers, poly(A) is at least part of the signal in question.

3. Splicing

Our understanding of the mechanism whereby introns (intervening sequences) are removed from messenger precursors has advanced dramatically over the past few years, largely because of three technical developments: the availability of antibodies against UsnRNP's, high-resolution electron microscope studies of disrupted nuclei, and most of all the development of in vitro systems in which introns are spliced efficiently and with high fidelity. This part of the field is in a period of rapid growth, and my comments here will certainly be out of date by the time this appears in print.

The development of in vitro systems by two groups during 1983-4 (Hernandez and Keller 1983; Padgett et al. 1983; Grabowski et al. 1984; Krainer et al. 1984, Padgett et al. 1984; Ruskin et al. 1984) rapidly led to the following findings. (a) Introns are excised as lariats (covalently closed branched circular RNA loops). This crucial discovery is illustrated schematically in Fig. 5 a. (b) The process has an absolute requirement for ATP, Mg and monovalent cations. (c) U1-snRNP, at least, is necessary. The complex of components responsible for splicing activity has since been purified about 100-fold (Di Maria et al. 1985); in this work, it appeared that GTP was another absolute requirement. Studies over this period also clarified the stringent sequence requirements at exon-intron junctions and provided evidence for a theoretical minimum intron length. Thus, by use of inherited splicing disorders such as analbuminemia and beta-thalassemia, and by site-directed mutagenesis, Esumi et al. (1983), Treisman et al. (1983) and Wieringa et al. (1983) contributed to the summary picture shown in Fig. 5 b. Wieringa et al. (1984), by artificially manipulating the large intron in the rabbit beta-globin gene, found that a 6-nucleotide 5'-proximal sequence and a 12-nucleotide 3'-proximal sequence were absolute requirements; full splicing efficiency was achieved with introns of length greater than 24 nucleotides; but a minimum of 80 nucleotides was necessary to ensure splicing fidelity. More recently, the structure of the lariat branch site itself has been definitively characterized by Sharp's group (Konarska et al. 1985).

a

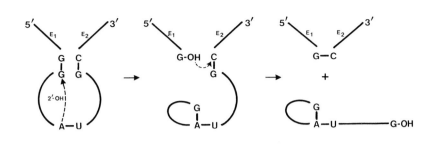

b

Fig. 5. The splicing mechanism. Scheme *(a)* shows the general mechanism by which introns are believed to be excised in the form of lariats, and *(b)* shows a generalized sequence that is apparently typical of intron/exon junctions *(py* pyrimidine; the *dots* represent individual bases that are not apparently conserved)

Before the in vitro systems were perfected, there were already indications that lariats appeared as splicing intermediates (Wallace and Edmonds 1983), and the trans-esterification step that is now believed to be involved (Fig. 5 a) had an established precedent in rRNA processing (Cech et al. 1981). Using Miller spreads of nuclei, Beyer et al. (1981) had shown that loops of RNA appeared in conjunction with the HnRNP particles. It is now believed that these apparently protein-free loops are lariats. This is consistent with the findings that introns are more RNAase-labile than exons (Steitz and Kamen 1981) and that the exon sequences, not the introns, are bound to the HnRNP core proteins (Munroe and Pederson 1981) (see Sect. C.II, below, for further discussion). The role of U1-snRNP was debated for some time before the development of the in vitro systems. Evidence for its involvement included (a) sequence complementarity between U1-snRNA and the conserved exon-intron junction sequences ("consensus sequences"), (b) inhibition of splicing by specific antibodies against U1-snRNP, which were mainly obtained from systemic lupus erythomatosus patients (Lerner and Steitz 1979; Lerner et al. 1980; Epstein et al. 1982; Mount et al. 1983; Kraemer et al. 1984; Grabowski et al. 1984). Before 1984, some workers found U1-snRNP involvement difficult to accept, possibly because U1-snRNP itself was not fully characterized. It is now possible to reconstitute this RNP particle in vitro (cf. Wieben et al. 1983), but the nature of its interaction with HnRNP, involving protein-

protein, RNA-RNA and protein-RNA linkages, is still far from clear (Harmon et al. 1985). Other UsnRNP's might be involved in splicing. For instance, there seems to be good evidence that U7-snRNP is required for processing of histone mRNA precursors (Strub et al. 1984).

The biological significance of splicing seems to lie in the capacity that it provides for alternative RNA processing. By varying the splicing activity so as to combine exons in various ways, the cell can obtain many messengers, hence many variants of a protein, from one gene. Processing of the calcitonin transcript is a well-investigated example (Rosenfeld et al. 1984), and other cases include immunoglobulins (Early et al. 1980), fibronectin (Schwartzbauer et al. 1983) and alpha-amylase (Young et al. 1981).

However, in this aspect of post-transcriptional processing, just as in capping and adenylation, one can argue for a role in mRNA transport. For the most part, RNA that is not completely spliced does not appear in the cytoplasm. There are apparent exceptions; for instance, SV40 late transcripts can pass from nucleus to cytoplasm in frog oocytes without complete splicing (Anderson et al. 1982; Wickens and Gurdon 1983), and the increased cytoplasmic messenger complexity early in carcinogenesis probably involves leakage of some immature precursors (cf. Shearer 1974, 1977). However, as a general rule, it seems that possession of one or more introns restricts RNA to the nucleus. This is obviously important for understanding the release stage of RNA transport. A particularly interesting result in this context was described by Gruss et al. (1979). They found that genes microinjected into nuclei were efficiently transcribed, capped and adenylated irrespective of the number of introns that had been left in them. The introns were duly spliced; but unless at least one intron had been left in the gene, the mature messenger would not enter the cytoplasm. The implication here is that nucleocytoplasmic mRNA transport can occur only if there are introns to be excised, and only after they have been excised. However, some viral genes are intron-free, and their transcripts are rapidly and efficiently transported to the cytoplasm (Mariman et al. 1982).

4. Messenger Stability and Degradation

I have mentioned the evidence that 5' caps and poly(A) tails protect messengers against degradation. Since messenger breakdown is important for exerting fine control, e.g. by hormones, over the specialized activities of highly differentiated cells (see Sect. E), it is important to consider this aspect of mRNA metabolism. It is possible that the topic has a direct bearing on the cytoskeletal binding aspect of mRNA transport. As things stand at the moment, our understanding of the mechanisms of mRNA degradation can best be seen as an extension of our knowledge of poly(A) metabolism. However, this is certainly not the whole story. Krowczynska et al. (1985) found that when erythroid transformation was induced with dimethylsulphoxide, many non-globin messengers, including those for actin and tubulin, were destabilized, and the stability showed no simple correlation with poly(A) length. It probably did not correlate with cap integrity either, because stability did not correlate with translational efficiency.

Mueller and his colleagues have identified and characterized a number of poly(A) catabolizing enzymes and have further characterized the apparently unique anabolic enzyme that is responsible for synthesizing the poly(A) tail. The poly(A) polymerase has been isolated from both nucleus (Edmonds and Winters 1976) and cytoplasm (Tsiapalis et al. 1975) of higher eukaryotes. Its activity can be regulated by protein phosphorylation (Rose and Jacob 1980; Schroeder et al. 1983). Three distinct catabolic enzymes have been identified. Endoribonuclease IV (Mueller 1976) is a 45 kDa protein, first isolated from chick oviduct, located in the nucleus (possibly also in the cytoplasm), which is highly specific for poly(A) which it cleaves into oligonucleotides with minimum lengths of around ten bases. Endoribonuclease V (Schroeder et al. 1980b) is a 52 kDa protein, first isolated from calf thyms, which cleaves poly(A) to 3'-AMP and poly(U) to 3'-UMP. Schroeder et al. (1980b) also described 2',3' exoribonuclease which cleaves oligo(A) or poly(A), alone or complexed with poly(U), to 5'-AMP. The studies by this team have indicated that poly(A) tail length is determined by the tuned interrelationships between these four enzymes. Thus, catabolic activities increase relative to anabolic activities during ageing, and decrease for specific messengers in response to hormonal signals. It is possibly by these means that messenger stabilities are altered during ageing and during hormonal stimulation of cells (see Sect. E).

Although the details of the mechanisms whereby stabilities of individual messengers are controlled are not yet clear, it seems well-established that certain messenger-associated proteins modulate the activities of the catabolic enzymes. Most particularly, the major poly(A) associated protein, which has a molecular weight of 75–78 kDa, and a 52–54 kDa protein, inhibit endoribonuclease IV and 2',3'-exoribonuclease activities when they are bound to adenylated messengers (Mueller et al. 1978). These proteins will be discussed further in the next subsection. In the meantime, it should be noted that the affinities of these proteins for poly(A) tails and other sensitive regions of messengers could, in effect, be the primary determinants of stability.

II. Aspects of Ribonucleoprotein Structure

It is well-known that cytoplasmic mRNA is specifically associated with a small set of polypeptides, and that the same is true of its intranuclear precursors. The sets of associated polypeptides in cytoplasm and nucleus are different (Dubochet et al. 1973; Kumar and Pederson 1975). HnRNP is now well-characterized, and in the first part of this section, I shall summarize some of the main stages in this characterization to date. The history has been one of relating detailed biochemistry to ultrastructure, and in this respect it resembles the elucidation of nucleosome structure (cf. Klug 1983). Messenger RNP is characterized up to a point, but some uncertainties remain. I shall describe the progress of this part of our understanding in the second part of this section.

1. HnRNP

The first indication that nuclear pre-mRNA was complexed with proteins to form a large ribonucleoprotein particle (40S) was given by Samarina et al. (1968). Despite some initial skepticism, this concept gradually gained acceptance. In the 1970s, several attempts were made to determine which proteins were tightly associated with HnRNA and which ones became artefactually associated during manipulations of the material (see e.g. Pederson 1974). This work showed the importance of maintaining RNA integrity during isolation of RNP. The first clear evidence for a major sextet of proteins tightly associated with HnRNA was given by LeStourgeon and his colleagues (Beyer et al. 1977). The sextet comprised three groups of proteins, which these authors named A1 and A2 (33–35 kDa), B1 and B2 (36–39 kDa) and C1 and C2 (41–43 kDa). Subsequently, this view of HnRNP structure has been repeatedly confirmed, for instance by photoaffinity labelling in situ (Economidis and Pederson 1983a). However, it is possible to obtain much more complex sets of proteins more or less tightly associated with HnRNA, and since these sets appear to be fairly reproducible, it may be appropriate to think of a sextet of "core" proteins and a less central set of "associated" proteins (see e.g. Stevenin et al. 1977; Dreyfuss et al. 1984). The term "core" is appropriate because the topology of the particle seems to be equivalent to that of the nucleosome: the nucleic acid is wound around a central complex in which the proteins are in intimate contact with one another, as cross-linking results indicate (Martin et al. 1978). Some of the "associated" proteins could be considered parts of the putative "nuclear matrix" (see next subsection). However, the expression "HnRNP particle" is taken to refer to HnRNA plus the "core" sextet.

   Although there seems to be no particular sequence specificity in HnRNA binding to the "core" proteins (though see Van Eekelen and Van Venrooij 1981), there is a good deal of evidence that the association involves exons, not introns. Some evidence for this was given in the discussion of splicing (previous subsection). Other lines of evidence, largely based on photoaffinity labelling, include the association of the "core" sextet with beta-globin messenger sequences in reticulocyte nuclei (Pederson and Davis 1980), with ovalbumin messenger sequences in oviduct nuclei (Thomas et al. 1981), and with actively transcribing polyoma virus sequences (Steitz and Kamen 1981). Also, Munroe and Pederson (1981) found that messenger sequences were better protected (by associated proteins) against RNAases than were intervening sequences.

   Ultrastructural evidence based on Miller spreads of nuclei have corroborated this conclusion. Beyer et al. (1980), for instance, found that the arrangement of HnRNP particles in such preparations was sequence-dependent, not random. Excision of introns occurs between the particles. The particles, of 25 nm diameter, are seen at the bases of loops comprising smooth 5 nm fibrils, apparently nascent lariats (Beyer et al. 1981; see previous subsection). However, it should be noted that the sudden exposure of nuclei to low ionic strength conditions, which is the essence of the spread technique, is likely to produce artefacts which are, in their way, potentially as misleading as those produced by high-salt treatment (Sommerville 1981; Tsanev and Djondjurov 1982).

Nevertheless, not only are ultrastructural descriptions an important aspect of HnRNP characterization, they have also been crucial for another equally important aspect: in vitro reconstitution. Particles apparently corresponding ultrastructurally, as well as biochemically, to the in situ HnRNP particles have been generated from HnRNA and the sextet of "core" proteins (Economidis and Pederson 1983b; Pullman and Martin 1983). In the elegant study by Economidis and Pederson (1983b), an adenovirus restriction fragment was transcribed with RNA polymerase II and the transcript formed an HnRNP particle with the protein sextet. Not only was poly(A) unnecessary for particle formation, but also particle formation did not seem necessarily to lead to splicing. It was in the light of this information that Pederson (1983) discussed the relevance of particle structure to processing.

2. mRNP

Methodological difficulties in analyzing messenger ribonucleoprotein complexes are rather similar to those that have faced workers analyzing HnRNP. Until the photoaffinity labelling technique was established, it was very difficult to distinguish genuine in situ-associated proteins from those from the cytoplasm or the cytoskeleton that became adventitiously associated during isolation. Like HnRNP, and unlike nucleosomes, mRNP is labile. Much use has been made of mRNP binding to oligo(dT) columns, but the procedures used in binding and fractionating are harsh enough to cause artefacts. Finally, methodologically sound distinctions between polysomal and "free" mRNP have been difficult to achieve. Some of these problems still persist, and there is as yet no ultrastructural characterization of mRNP to compare with that of HnRNP. Despite these problems, some progress has been made and substantive conclusions can be drawn from the existing information.

During the 1960s, Spirin and his collaborators obtained evidence that mRNA was protein-associated in situ (see e.g. Spirin 1969). In the 1970's, attempts were made to characterize free mRNP by taking advantage of its density-gradient behaviour, its somewhat greater sensitivity (compared with polysomal mRNP) to RNAase, and its affinity for poly(U) or oligo(dT) columns. The starting-point for studies on polysomal mRNP was the discovery that it could be dissociated from ribosomes with EDTA (e.g. Perry and Kelley 1968). Most studies on both sources of mRNP agreed in respect of two apparently ubiquitous protein components, with molecular weights of 72–78 kDa and 48–52 kDa (see e.g. Morel et al. 1971; Blobel 1973; Bryan and Hayashi 1973; Barrieux et al. 1975; Kumar and Pederson 1975), though quite different, and often more complex, sets of polypeptides were also reported (e.g. Gander et al. 1973; Bag and Sarkar 1976).

Two groups of proteins have been definitively located on specific regions of mRNA. One of these groups, the cap-binding proteins, was discussed in the previous subsection. The other is the 74–78 kDa polypeptide (P78). The early evidence that this is associated apparently exclusively with poly(A) (Blobel 1973; Barrieux et al. 1975; Schwartz and Darnell 1976; Jeffery 1977) has been amply corroborated, not least by the photoaffinity labelling studies of Van Eekelen et al. (1981). There is also some evidence that the other ubiquitous protein, the 48–52 kDa component (P52), is associated with poly(A) (Mueller et al. 1985). No doubt these associated proteins are important in protecting poly(A) against the catabolic enzymes mentioned at the end of the previous subsection, and thus in regulating mRNA stability.

An interesting and potentially important recent development is the reversible assembly/disassembly of mRNP in vitro (Gaedigk et al. 1985). However, apart from P78 and P52, the particles assembled by these workers contained several other proteins (98, 46, 42, 40, 34, 28 and 24 kDa) of cytoplasmic origin. It would be interesting to know whether any of these proteins was associated with the cytoskeleton. The other most interesting recent work is that of Schweiger and his colleagues, who have identified a poly(A)-associated protein of 110 kDa which is apparently found in both nucleus and cytoplasm (e. g. Schweiger and Kostka 1984). This protein, or one of very similar molecular weight and apparent poly(A) affinity, will be discussed further in Section D.

So far as the relevance of RNP structures to mRNA transport is concerned, the main point is this. Prior to the release stage of transport, mRNA (or its precursor) is associated with one set of proteins; after cytoskeletal binding, it is associated with another. Which proteins, if any, remain associated with the messenger during translocation? There is no answer to this question as yet, but some results from the study of the translocation mechanism (see Sect. D), and the work of Van Eekelen et al. (1981), suggest that P78, at least, is not associated: it is restricted to the cytoplasm. Attempts to answer this central question satisfactorily could well direct the efforts of many researchers in the field over the next few years.

III. Relationships to Subcellular Structure

1. The Nuclear Matrix Controversy

If the solid-state view of mRNA transport is valid, then intranuclear messenger precursors are immobilized on a structure that ultimately presents them to the pore-complex and releases them for translocation (cf. Fig. 1 c). Therefore, the "nuclear matrix" controversy — the argument about whether such a structure exists in vivo and, if it exists, what its characteristics are — is centrally important to the study of mRNA transport. The literature on the nuclear matrix is enormous, so it cannot be reviewed fully here. I shall concentrate on what seem to me the fundamental parts of the controversy, especially as they relate to mRNA transport. For reviews of earlier work, and of the traditional arguments in favour of the reality of the matrix, see Wunderlich et al. 1976; Berezney 1979; Agutter and Richardson 1980. Most authors of papers in the field accept the existence of an intranuclear structure, or assume it uncritically, or simply adopt an operational definition of "nuclear matrix"; but some give compelling arguments. The critics are relatively few, but for the most part their experiments are better-conceived and their arguments are more analytically penetrating.

I shall discuss the issue under the following headings:
a) What is the basic ontological argument really about?
b) What are the critics really criticizing?
c) In the light of this, what arguments can the protagonists of the matrix concept legitimately maintain?
d) How can we explain the apparent inconsistencies in the field?
e) What future directions can we take in the analysis of HnRNP interactions with the putative nuclear matrix?

a) The Ontological Argument

The question is not simply: "does the nuclear matrix exist or is it an artefact?" There are at least four possible views:

1. The in vivo nucleus contains chromatin and HnRNP but no other multimolecular structure. The rest of the intranuclear space contains freely diffusible proteins in aqueous solution.
2. It is as "crowded" as the cytoplasm (cf. Fulton 1982), so that much of the protein is in a gel state.
3. It is organized by a morphologically heterogeneous structure (cf. Berezney and Coffey 1974; Berezney 1980) akin to the proposed "microtrabecular lattice" of the cytoplasm (Wolosevick and Porter 1976).
4. It contains a delicate fibrillar system (Brasch 1982; Capco et al. 1982; Fey et al. 1984), akin to the intermediate filaments of the cytoskeleton.

Powerful evidence that only about 30% of the nucleus contains free water (Lang and Peters 1984) might throw doubt on the first of these four possibilities, but the others remain open. Of these, possibility (2) is quite useless in terms of the solid-state perspective: it does not meet the basic mRNA transport requirement set out above. Its validity (or otherwise) is therefore not relevant to the discussion in this review. The only possibilities compatible with the solid-state perspective are (3) and (4).

b) Targets of Criticism

1. Protein cross-linking can be caused by glutaraldehyde fixation of specimens for electron microscopy (Skaer and Whytock 1977) and traces of calcium can aggregate and precipitate soluble nuclear components (Lebkowski and Laemmli 1982a, b; Comerford et al. 1985). Some intranuclear proteins can cross-link and form artefactual fibrils by spontaneous or chemically induced oxidation of endogenous sulphhydryls

[Kaufmann etal. 1981; see (d), below]. Obviously, such artefacts would generate structures with a heterogeneous, complex composition, which in fact most isolated "nuclear matrix" preparations have.

2. Salt-induced collapse and aggregation of regions of the nucleoplasm containing high chromatin concentrations (Okada and Comings 1980; Hadlaczky etal. 1981; Burkholder 1983) can also produce artefactual "nuclear matrices". Of course, these could contain enzymes such as DNA polymerase, and specific hormone binding sites, and infecting virus particles, not to mention newly replicated DNA and some or all of the HnRNP. Thus, the favourite arguments of matrix protagonists, i.e. that their preparations contain just such specific markers and must therefore represent specific nuclear subfractions, become indecisive and possibly vacuous.

3. More particularly, some very careful studies have shown that single-stranded, including newly replicated, DNA has a particularly high affinity for such structures in the presence of magnesium ions and high salt concentrations (Ross etal. 1982; Krachmarov etal. 1986). The implication that there is no sequence-specific DNA association with these "nuclear matrices" received spectacular support from the finding by Forbes etal. (1983) that any DNA, including bacteriophage DNA, would induce the formation of morphologically normal nuclear envelopes when injected into oocyte cytoplasm.

4. Lestourgeon and his colleagues (see e.g. Arenstorf etal. 1984) have similarly shown that HnRNP proteins, specifically the A and B groups, artefactually reassociate to form fibrils when exposed to high salt concentrations. These fibrils have a fairly regular diameter and helical pitch, and become stabilized by disulfide linkages, after nuclease digestion. They have a recognizable core unit, typically a tetramer comprising three molecules of A2 and 1 of B1, but many other proteins can then associate with them in high salt.

Almost all these arguments throw very serious doubt on the in situ reality of "nuclear matrices" of the Berezney type [see possibility (3) under (a), above]. They do not exclude the idea of an intranuclear gel or of a delicate fibrillar system [possibilities (2) and (4) in (a)], a point on which Krachmarov etal. (1986) are explicit. Most of all, they do not "prove" that the intranuclear space is structured only by chromatin and HnRNP [(1) in (a)].

c) Surviving Arguments in Favour of the Matrix Concept

1. Cook and Brazell (1975) showed that "nucleoids" (i.e. nuclei expanded by 2M NaCl treatment, histone-free but containing total intact DNA as a spread "halo") had a sedimentation rate that responded biphasically to increasing ethidium bromide intercalation, i.e. their DNA behaved as circular, supercoiled DNA. This response was progressively abolished by X-irradiation, and target-size analysis in these experiments led the authors to conclude that the DNA was organized to form supercoiled "loops" of 80–100 kbp. Similar values for the "loop" size were obtained by Vogelstein etal. (1980) on the basis of fluorescence-microscope measurements of nucleoid diameters in various ethidium bromide concentrations; and by Igo-Kemener and Zachau (1978)

on the basis of progressive nuclease digestion of intact nuclei. Here, salt extraction is evidently not responsible for all the observations. Moreover, metaphase chromosomes exhibit a comparable structure with similarly sized DNA loops when examined in section (Marsden and Laemmli 1979) or by scanning electron microscopy (Adolph and Kreisman 1983; see also Mullinger and Johnson 1980), suggesting that interphase and metaphase chromosomes might be analogously structured. The existence of such large loops presumably implies the existence of a large-scale anchoring structure in the nucleus, and this argument seems immune to the evidence provided by the critics (Gooderham and Jeppeson 1983). Also, the interphase matrix might correspond to the metaphase chromosome "scaffold" or "core" (Adolph et al. 1977; Jeppesen et al. 1978; Laemmli et al. 1978; see also Stubblefield 1973).

2. A dense, anastomosing reticular system of fine fibrils has been identified in nuclei in situ, both in thin sections of e.g. alpha-amanitin-treated liver in which the chromatin has contracted (Ghosh et al. 1978; Brasch 1982; Pouchelet et al. 1984), and in thick sections of untreated cells viewed by high-energy transmission electron microscopy (Capco et al. 1982; Granger and Lazarides 1982; Guatelli et al. 1982; Capco and Penman 1983; Fey et al. 1984). The latter studies suggest that this framework is continuous through the pore-complexes with intermediate filaments. The immunological similarity between the lamins and cytokeratins (Franklin et al. 1984) perhaps supports this suggestion.

3. A small number of possible "core" polypeptides have been identified in a fibrillar system isolated from interphase nuclei and metaphase chromosome "scaffolds" by nuclease/salt extraction (Lebkowski and Laemmli 1982a, b; Pieck et al. 1985) and by alternative methods (Mirkovich et al. 1984). According to Laemmli and his colleagues, the fibrillar structures in both the interphase and the mitotic cell are cuproprotein in nature. In *Drosophila* nuclear matrices, the small set of core polypeptides seems to be dominated by DNA topoisomerase II, which is particularly interesting in view of point (1), above (Fisher et al. 1982; Berrios et al. 1983; Berrios et al. 1985). Structures comprising a small set of dominant polypeptides are unlikely to be artefacts caused by random precipitation of nuclear contents.

4. Monoclonal antibodies have been produced that label a fibrillar system in nuclei (Chaly et al. 1983, 1984; Bhorjee et al. 1983) and in metaphase chromosome "scaffolds" (Sternberg 1984) in situ. Polyclonal antibodies, either raised in mice or rabbits or obtained from sera of human patients with autoimmune diseases, have been deployed to similar effect (Krohne et al. 1982; Wojtkowiak et al. 1982; Salden et al. 1982; Habets et al. 1983a, b; Fritzler et al. 1984).

5. Very interestingly in the context of this review, an intranuclear actin filament has been identified (Clark and Rosenbaum 1979) that seems to bind snRNPs in situ (Nakayasu and Ueda 1984).

6. RNA is moved fairly rapidly from the sites of transcription and processing, and is directed specifically to the pore-complexes. The distances involved are in the order of microns. It is difficult to see how this could be achieved in the absence of a fibrillar communication system of some kind.

It will be noted that the arguments in favour of the matrix concept actually favour the existence of a delicate, cytoskeleton-like framework in nuclei in situ, not a morphologically heterogeneous structure. That is, they are compatible with possibility (4), but not so much with possibility (3), in (a). On this analysis, it is both surprising and reassuring to find that the conclusions of the critics and of the most recent protagonists do not, after all, conflict seriously [cf. final paragraph in (b), above]. However, there are some apparent inconsistencies in the literature.

d) Resolutions of Some Inconsistencies

Contemporaneously with the publication of the first procedure for isolating nuclear matrices from rat liver (Berezney and Coffey 1974), Aaronson and Blobel (1975) described the isolation of the pore-complex lamina. Not only was the source of nuclei the same in the two laboratories involved (rat liver), but also the steps in the isolation procedure were apparently more or less identical (salt-extraction, nuclease treatment and Triton X-100 extraction of the nuclei). The protein compositions of the two preparations seemed to be dominated by the same polypeptide triplet (comigrating with lamins), and the immunological localization of the lamins at the nuclear periphery (Gerace et al. 1978) cast further doubt on the in situ reality of the matrix.

Why should apparently similar techniques applied to identical starting materials generate such strikingly different structures? This question was addressed by Kaufmann et al. (1981), who in a careful and illuminating study demonstrated that a variety of morphologically distinct structures can be isolated from the same batch of nuclei by subtle alterations in the operation times and in the order in which the steps in an isolation procedure are applied. Their main findings were as follows. (1) If protein sulphhydryl group were reduced, only the residual envelope was isolated after salt and nuclease extraction; if oxidation was allowed so that intermolecular disulphide bonds formed, then intranuclear structures were isolated. (2) If ribonuclease treatment preceded salt extraction, no intranuclear material was isolated; but if salt extraction preceded RNAase treatment, intranuclear fibrils were found in the isolated nuclear residue. The first of these findings simply suggested that the putative intranuclear fibrils are stabilized by intermolecular disulphide bridges [see under (b), above], and it explained the conflict between the two 1974 publications. Aaronson and Blobel (1975) had isolated nuclear envelopes by the Kay et al. (1972) procedure, which involves the use of 2-mercaptoethanol. Berezney and Coffey (1974) had used much longer operation times and had taken no steps to prevent oxidation.

The second main finding of this study is less readily interpretable. However, suppose that there is, in situ, a delicate intranuclear fibrillar framework and that this is more stable when HnRNP (+ snRNP) is bound to it than when it is RNP-free. Given that salt extraction of chromatin is a violently disruptive procedure at the molecular level, such a framework could survive salt extraction only if was stabilized, i.e. if RNP remained bound to it. If it survived, then other nuclear components (proteins, nascent DNA, etc) could associate with it artefactually, producing the morphologically heterogeneous structures previously described. This view is consistent with the conclusions in (b) and (c), above, and will be discussed further in the next paragraphs. The absence of any intranuclear structure after extraction of chromatin with EDTA and DNAase II might be explained similarly (e. g. Krachmarov et al. 1986).

e) Directions for Future Progress

First, a small philosophical point. The structures that have been considered in this chapter — the various elements of the cytoskeleton, the lamina, HnRNP — are *well-characterized* in the following senses. First, they have defined, regular, reproducible morphologies, identical in situ and in isolated preparations. Second, their principal polypeptide components have been identified. Third, they have been labelled by specific antibodies, which have been shown to bind with specific individual components. Forth, except for the lamina, they have been reversibly disassembled/assembled in vitro. Further characterization is in progress in these areas: issues of detailed molecular architecture and the biochemical basis of the dynamics of the structures are being investigated. (Incidentally, only the first aspect of characterization has been met, albeit very thoroughly, for the pore-complex. The consequence is the proliferation of models for this structure (cf. Fig. 3).

As far as the matrix is concerned, the heterogeneous structure [(3) in (a), above] has not only been cast into doubt by the critics, but has not been well-characterized in any respect. However, the intranuclear fibrillar system is on the way to meeting at least the first three aspects of characterization, which would give it the same ontological status as the lamina.

The obviously pressing need is for an isolated preparation that has the dense reticular ultrastructure of the fibrillar system visualized in situ, contains the core polypeptides identified by Laemmli's group (along, presumably, with lamins and HnRNA-associated proteins) and is labelled by the specific antibodies that are successfully applied to the intact cell. We have obtained a preparation from liver nuclei that has the appropriate ultrastructure and the appropriate polypeptide composition, and our preliminary results show assembly/disassembly of the fibrils in vitro by addition and withdrawal of 0.1 μM copper. Most interestingly, RNAase treatment of the isolated matrix disperses the reticular network but leaves the individual, separate fibrils intact (see Fig. 6), and this is associated with solubilization of some of the protein components, but not the core polypeptides described by Lebkowski and Laemmli (1982 a, b): these latter are solubilized only with copper chelators. Cytochalasin B simulates the effect of RNAase exactly (Fig. 6). It will be noted that the RNAase effect is precisely what would be predicted from the hypothesis advanced at the end of (d), above, on the basis of the studies by Kaufmann et al. (1981). Also, the cytochalasin B effect is immediately consistent with the concept of a RNP-binding actin network in the nucleus (Nakayasu and Ueda 1984).

These findings are very promising, but appropriate work with monoclonal and other antibodies, further polypeptide analysis and more definitive assembly/disassem-

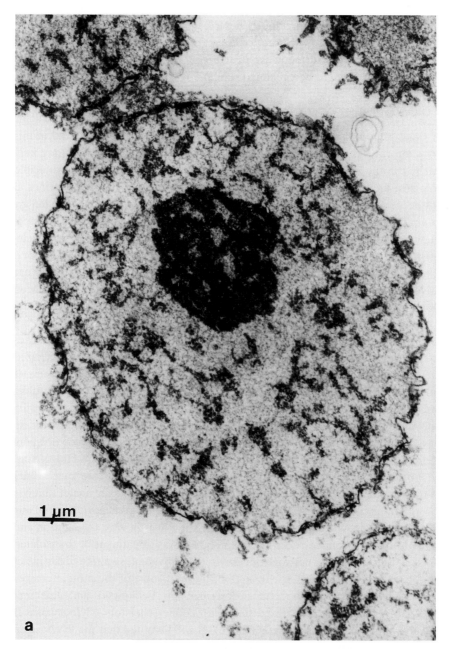

Fig. 6. Electron micrographs of a dense, fibrillar reticular structure isolated from rat liver nuclei by the method of *(a)*, *(b)* untreated preparations; *(c)*, preparation after RNAase treatment; *(d)*, preparation after cytochalasin B treatment

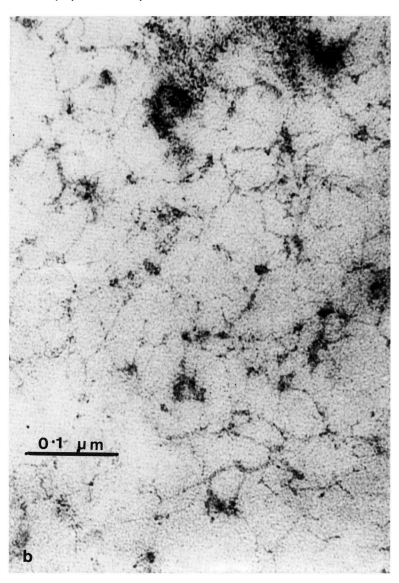

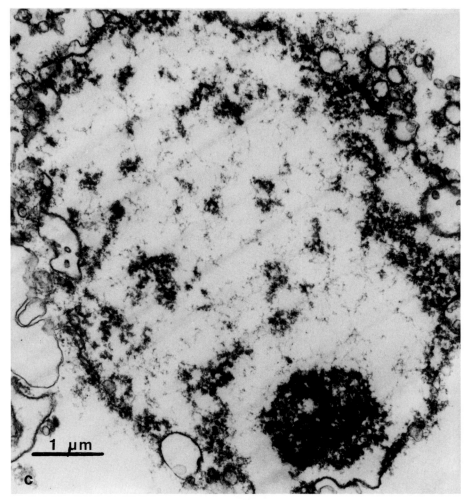

Fig. 6 c,d.

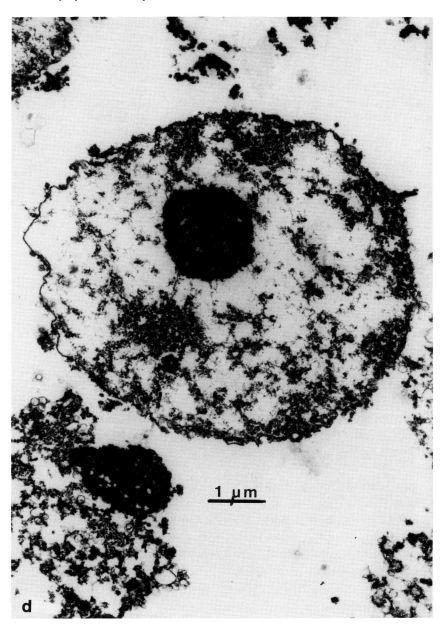

1 µm

d

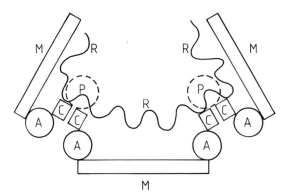

Fig. 7. Possible scheme for the organization of HnRNP and the nuclear reticulum. The short reticular fibrils *(M)* that survive RNAase or cytochalasin B treatment (see Fig. 6) terminate in junctions with short actin bundles *(A)*. These in turn are linked to the C-group proteins *(C)* of HnRNP core particles. The HnRNA itself is indicated by the *undulating line (R)* and the *A* and *B* group proteins of the core particle by *(P)*. It should be emphasized that this scheme is highly speculative

bly studies are needed. In the meantime, some speculations can be made about the molecular architecture of the fibrillar matrix, incorporating ideas about HnRNP-fibril interaction that are relevant to the release stage of mRNA transport. One speculation is shown here (Fig. 7), and it is based on the information described above and on the photoaffinity-labelling evidence that HnRNA is attached to the matrix via the C-group proteins of the HnRNP complex, to which it is attached by oligopyrimidinosine sequences (Van Eekelen and Van Venrooij 1981; Van Eekelen et al. 1981). The C-group proteins could be the antigen common to HnRNP and matrices (Gallinaro et al. 1983). It should be noted that in some systems, at least, the RNA might bind directly to the matrix fibrils, not through the intermediary of any of the major HnRNP proteins. Maundrell et al. (1981) described this situation in duck erythroblasts. These authors said that the globin coding sequences were involved in the binding (but see Ross et al. 1982).

Such schemes should become testable when the isolated preparations are better characterized. However, there is another direction that future studies could usefully take in parallel with such further characterization. The polypeptide components of the nuclear matrix must cross-link more readily with each other than with non-matrix components. In situ cross-linking studies could therefore give a more satisfactory idea about the protein composition of the matrix.

One final word on the subject: perhaps we should drop the use of the term "matrix" in view of its ambiguity [cf. (a), above], and introduce another phrase, such as "nuclear skeleton" or "nuclear reticulum", to describe the delicate intranuclear fibrillar structure in which we can now legitimately believe.

2. Messenger Ribonucleoprotein and the Cytoskeleton

The existence of microtubules, microfilaments and intermediate filaments in cytoplasms of eukaryotic cells is now universally accepted. There is no controversy about the reality of these structures as there is, or has been, about the nuclear matrix. However, it is worth remembering that there was just the same sort of skepticism about the cytoskeleton until late in the 1960s, and even today some issues remain unclear. For instance, the precise nature and roles of microtubule-associated proteins, the detailed molecular architecture of the fibrillar elements of the cytoskeleton, and the molecular basis of microtubule and microfilament dynamics, are all active areas of research and debate.

One of the unclear areas is the mode of association of mRNP with the cytoskeleton. Little has been done so far to address the topic in a systematic way. The main questions to be asked would seem to be these:

a) To what element(s) of the cytoskeleton is mRNA bound, and by means of which protein components?
b) What part of the polysome binds?
c) Why is it that cytoskeletal binding of polysomes seems to be a prerequisite for translation?

As was mentioned earlier in this chapter, the likeliest answer to (a) is that messengers bind to intermediate filaments. They are close to the nucleus, apparently continuous with the "traverse fibrils" of the pore complexes, in some cases (cytokeratins) biochemically similar to the lamins, and at least in one case tightly associated with a cap-binding protein. However, it is not possible to exclude functionally important associations of mRNP with other cytoskeletal elements in the present state of knowledge.

So far as question (b) is concerned, it is at least clear that polysomes bind to the cytoskeleton through the messenger itself, not through the ribosomes. Van Venrooij et al. (1981) were able to remove the ribosomes reversibly from their detergent-extracted cell residues without detaching the messengers. However, there is no evidence for sequence specificity in the binding as yet. One reasonable possibility is that one or more of the more tightly bound messenger-associated proteins plays the key role in cytoskeletal binding. This would suggest P78 (cf. Mueller et al. 1985), or of course the cap-binding protein itself. Against the former possibility one could draw attention to the lack of correlation between poly(A) tail length and translatability, and to the absence of poly(A), and hence presumably of associated P78, from most histone messengers. In favour of both possibilities, however, is the logical point that cytoskeletal binding is likely to involve mRNA sequences that are common to all or most messengers, such as the 3'-poly(A) tail and the 5'-methyl cap. If, indeed, cytoskeletal binding is effected through the cap and the cap-binding protein(s), then this might, in outline, suggest an answer to question (c).

In this area, definitive information is scarce and further speculation is pointless. There is a need for research groups with expertise in manipulating the cytoskeleton to collaborate with others whose expertise lies in manipulating mRNP. Only through such a multidisciplinary approach will this aspect of mRNA transport be significantly elucidated.

D. Messenger RNA Transport

So far, I have discussed some recent progress in our knowledge of nucleocytoplasmic exchange processes, HnRNA and mRNA metabolism, RNP structures, nuclear envelope structure, the "nuclear matrix", and the methodological adequacy of in vitro systems for studying mRNA efflux. I have made a number of points relevant to mRNA transport, and it will probably be helpful to summarize these at this stage:

a) mRNA transport seems to be a solid-state process. That is, any HnRNA or mRNA that is *not* immobilized in situ is not physiologically significant (see Sect.A.III).

b) mRNA moves vectorially from nucleus to cytoplasm through the pore-complexes. It seems likely that fibrils continuous with both faces of these structures are important for transport. In fact, the pore-complexes might be structures that are moving along a continuous nucleocytoplasmic fibrillar system, possibly carrying mRNA and other materials with them, (see Sect.B.I).

c) Over nuclear envelope components that are likely to be involved are the lamins (especially lamin B, which is intimately associated with the inner nuclear membrane) and the inner membrane itself, which contains potential regulators of translocation.

d) Transport probably depends on both "permeation" and "association" signals in the transport substrate itself. For mRNA, the "permeation" signals (which facilitate movement) probably include poly(A), but there are certainly others. Nuclear "association" signals (which prevent egress to the cytoplasm) probably include the HnRNP core proteins and the introns. Cytoplasmic ones might include caps and the associated cap-binding proteins, and P78. However, signals of all kinds might depend on secondary-structure characteristics, or quaternary features of specific RNA-protein complexes, rather than simple sequences (see Sects. B.II, C.I and C.II).

e) Despite the continuing controversy about the nuclear matrix, it seems clear that nuclei contain a delicate, anastomosing, reticular fibrillar system in vivo, and that this is HnRNA-associated. Many isolated "matrix" preparations, however, are certainly artefacts, and the probable dependence of reticulum integrity on HnRNA integrity can generate misleading results (e.g. Bouvier et al. 1985) (see Sect. C.III).

f) Although there are no valid in vitro models for studying most nucleocytoplasmic exchanges, mRNA transport is exceptional. In vitro systems are now available that, within limits, are very valuable for elucidating mechanisms. The limitations are (1) efflux measurements on low-abundance messengers are difficult to interpret, (2) efflux does not model the cytoskeletal binding stage of transport (see Sect. B.III).

I made the point early in this chapter (Sect. A.II) that mRNA transport needs to be envisaged minimally as a three-stage process, comprising release, translocation and cytoskeletal binding steps.

In this section, I shall reflect further on our level of understanding of these three stages, and then proceed with a fuller discussion of the only one of the three, translocation, of which we have any detailed knowledge at present.

I. Stages in Transport

1. Release

If the intranuclear RNA is attached to a nuclear reticulum or "matrix", then detailed characterization of this structure is a prerequisite for understanding the release stage of mRNA transport. We do not yet have such detailed characterization, and therefore any claims about release are speculative at the moment. Nevertheless, some questions and comments can legitimately be raised.

a) Release is patently not a one-step event. First, the messenger (or precursor) has to be conveyed from an assembly site, which may be deep in the nucleus, before it is delivered to the pore-complex. Intranuclear movement and transfer to the pore-complex appear logically distinct, though they might not be *mechanistically* distinct, e.g. if attachment to the reticular fibrils operates through pore-complex precursors which then move bodily along the fibrils to the nuclear periphery, and perhaps beyond. Second, if both the HnRNP core proteins and the introns are nuclear accumulation signals, then since they are not directly associated with each other (see Sects. C.I and C.II), they have to be detached from the messenger in separate processes. Removal of mRNA from the core proteins is not an obvious consequence of the final splicing step. Thus, to speak of "release" as a single event is probably an oversimplification.

b) However, it is possible that the pyrimidine-rich sequences linked to the C-group proteins and hence, perhaps, to the nuclear reticulum are excised during splicing (cf. Van Eekelen et al. 1981, and Figs. 5b and 7).

c) Modifications of the HnRNP core proteins, other associated proteins, or the components of the reticulum itself might be necessary for release. In this context, studies on the phosphorylation of nuclear RNA-associated proteins might be relevant (Blanchard et al. 1975, 1977, 1978; Alonso et al. 1981).

d) Given that the reticulum probably contains actin, RNA movement might depend on contractile or "treadmilling" mechanisms (cf. Capco et al. 1982; Maul 1982; Capco and Penman 1983; Dazy et al. 1983; Agutter and Thomson 1984). Here, the possibility that the whole intracellular space is functionally organized by an integrated fibrillar system needs to be kept in mind (Cervera et al. 1981; Alekseev et al. 1982; Capco and Penman 1983; Murnane and Painter 1983). Also, reversible expansion and contraction of the "matrix" in nuclei from a lower eukaryote (*Tetrahymena* macronuclei) occur in response to changes in bivalent cation concentrations and temperature (Wunderlich and Herlan 1977; Herlan et al. 1980), and in this case RNA efflux seems to occur only when the matrix is expanded (Giese and Wunderlich 1982; Wunderlich et al. 1984b; Wunderlich et al. 1984a). It should be noted that Wunderlich and his colleagues have defined their "isolated matrix" operationally, and that their studies explicitly concern ribosomes, not mRNA. Whether such reversible expansion occurs in higher eukaryotic nuclei, and whether it is relevant to mRNA movement, remain open questions.

A more detailed understanding of release is desirable because the control of RNA transport is likely to be focussed predominantly on the release stage. It is the first stage in transport and the one most likely to determine the selection of messengers for export to the cytoplasm (i.e. nuclear restriction). There is no doubt that the control of mRNA transport is important in the fine control of gene expression and perhaps of cellular differentiation (see Sect. E). In the meantime, some models of mRNA transport have been proposed that ignore intranuclear events altogether (Clawson and Smuckler 1982a, b), and it may be instructive to consider the advantages and the limitations of these in the light of the discussion here.

2. Translocation

Our understanding of translocation is much more advanced than our knowledge of release, and indeed is sufficiently detailed to allow well-articulated kinetic models to be proposed (see Sect. D.III, below). Part of the reason for this is that reasonably satisfactory procedures for nuclear envelope isolation have been available for several years (contrast the state of affairs with regard to matrix isolation), and therefore the appropriate material for biochemical study of the translocation apparatus has been to hand. Another part of the reason may be that most workers have until fairly recently approached mRNA transport intuitively in terms of the "solution/diffusion" perspective, which regards translocation as the whole of transport and therefore necessarily focusses attention on this stage of the overall process.

The current weight of evidence indicates that translocation in at least many higher eukaryotic cells is energy-requiring and involves a nuclear envelope-associated nucleoside triphosphatase (NTPase, 3.6.1.15). This enzyme has been much studied. The involvements of protein kinases and phosphohydrolases, and of an RNA binding component in the envelope, are also strongly suspected. Current interest is focussed on purifying and characterizing these components. The detailed kinetic schemes that can currently be offered embody sets of hypotheses, the predictions of which can be tested by such characterization and by studies on the mechanisms of interaction of the components. I shall discuss these matters in detail in Sections D.II and D.III, below.

3. Cytoskeletal Binding

This topic has scarcely been studied at all at the biochemical level, and there is nothing to add to the remarks that I have made previously, viz: cytoskeletal binding occurs immediately after, and probably concomitantly with, egress of messengers from the pore-complexes; it is a precondition for translation; it probably involves intermediate filaments; and it requires some feature common to all messengers, such as the 5'-methyl cap + cap-binding protein or the 3'-poly(A) tail + P78. There is at least some evidence for the former possibility.

II. Translocation

1. The Nuclear Envelope Nucleoside Triphosphatase

The NTPase is an integral nuclear envelope component. It has a broad substrate specificity, hydrolyzing ATP, dATP and GTP (in the form of Mg or Mn complexes) at similar rates and CTP and UTP at about half that rate. Under some assay conditions it deviates from normal Michaelis-Menten kinetics (Agutter et al. 1979 a). It seems to be located on the inner face of the nuclear envelope, probably in the lamina or the nucleoplasmic face of the inner nuclear membrane (Vorbrodt and Maul 1980; Kondor-Koch et al. 1982). The evidence that it provides the energy for mRNA transport is as follows.

a) In isolated nuclear envelopes, the activity of the NTPase is markedly increased when exogenous RNA is added. Cytoplasmic mRNA and homopolymers such as poly(A) and poly(G) are effective stimulators. Ribosomal RNA, tRNA, poly(U), poly(C) and cytoplasmic mRNA from which the poly(A) tail has been removed or decreased to less than about 15 bases, are not effective stimulators. In order to be an effective stimulator, it appears that the polynucleotide must have an organized tertiary structure, though precise details of this requirement are not available (Agutter et al. 1977; Agutter and Ramsay 1979; Bernd et al. 1982a). These polynucleotides increase the catalytic constant of the NTPase reaction but do not alter the Michaelis constant of the enzyme for MgATP.

b) Nuclear envelopes can be made to form resealed vesicles with normal topology (Kondor-Koch et al. 1982). When mRNA is trapped in such vesicles it can be exported only in an apparently NTPase-dependent way (Riedel et al. 1987).

c) Conditions that result in an increased cytoplasmic/nuclear RNA ratio and increase the rate of mRNA transport in vivo increase the specific activity of the NTPase as measured in isolated envelopes. Examples are carcinogen feeding (Clawson et al. 1980 a, b), tryptophan feeding (Murty et al. 1980) and insulin treatment (Purrello et al. 1982) of rat liver. The NTPase activity is also age-related in a way that corresponds to the age-relatedness of the RNA transport rate (Bernd et al. 1982b, 1983).

d) The efflux of mRNA from isolated liver and other nuclei, under conditions in which efflux is a satisfactory model for the release + translocation stages of transport (see Sect. B.III), is nucleoside triphosphate-dependent. The rate of efflux shows the same substrate specificity as the NTPase activity does; it shows the same kinetics, with the same apparent Michaelis constant; it shows the same divalent cation specificity; and it shows the same sensitivity to inhibitors (Agutter et al. 1976, 1979b; Agutter

1980; Clawson et al. 1980 a). Conditions that alter the rate of mRNA transport in vivo also alter the rate of efflux (see previous paragraph for examples), and again this corresponds to changes in the NTPase activity.

e) A monoclonal antibody that inhibits RNA efflux also inhibits the NTPase. Monoclonals against other nuclear envelope components affect neither the NTPase nor the RNA efflux rate (Baglia and Maul 1983). The active monoclonal seems to react with a protein that comigrates with lamin B on SDS-gel electropherograms.

It is worth noting that two of these five lines of evidence (a, c) do not depend on RNA efflux studies.

2. Other Hypotheses About the Role of ATP

The hypothesis that RNA transport is energy-requiring and therefore depends on ATP hydrolysis was advanced by several laboratories long before there was adequate supporting evidence (Schneider 1959; Ishikawa et al. 1969; Horisberger and Amos 1970; Yu et al. 1972; Smuckler and Koplitz 1974, 1976; Kletzein 1975). It now seems to have been amply corroborated by evidence such as that summarized above. However, three other explanations of the ATP-dependence of RNA transport have been proposed.

a) The ATP simply functions as a chelating agent. This view is based solely on one set of efflux studies (Chatterjee and Weissbach 1973) that were performed using a medium that was demonstrably inadequate to support normal nuclear restriction — indeed, nuclei in this medium are lyzed by the addition of polynucleotides (Goidl et al. 1975), and this does not occur in more adequate media (Agutter 1980, 1983). In media in which RNA efflux corresponds to in vivo release + translocation, chelating agents such as EDTA have no measurable effect on the efflux rate.

b) The ATP causes expansion of the nuclear envelope, altering its permeability properties (Stuart et al. 1975, 1977). Although this hypothesis makes sense only in terms of the "solution/diffusion" perspective on mRNA transport, not in terms of the more adequate "solid-state" perspective, it is interesting. Many nuclear envelope polypeptides can be phosphorylated by ATP (Smith and Wells 1983; Agutter 1985c), and among these are the lamins. These are important structural components of the envelope and their phosphorylation is likely to alter the properties, including the permeability properties, of the system as a whole (Ottaviano and Gerace 1985) and seems to be causally related to the breakdown of the nuclear envelope at mitosis (Gerace and Blobel 1980, 1981). Isolated interphase rat liver envelopes contain lamins in various phosphorylation states (Kaufmann et al. 1983). Using resealed veşicles as described by Kondor-Koch et al. (1982) and colloidal gold permeation measurements, I have found evidence that ATP-dependent envelope phosphorylation increases the patent pore diameter by around 10% (unpubl. observ.), and similar results seem to have been obtained elsewhere (M. Schindler, pers. commun.). Thus, although ATP-induced nuclear envelope expansion is perhaps not directly related to RNA efflux and transport, it may be related to the exchanges of other materials across the envelope, including factors that are important in regulating mRNA transport itself. The importance of endogenous protein phosphorylation is quite another matter, and is discussed below.

c) The ATP causes displacement of mRNA from a nuclear envelope binding site by a mechanism analogous to allosteric modification (Raskas and Rho 1973). This is perhaps the most interesting of the alternative hypotheses about the role of ATP. Although Yu et al. (1972) and Agutter (1980) found that the non-hydrolyzable ATP analogue $\beta\gamma$-methylene-ATP had no significant effect on mRNA efflux, Ishikawa et al. (1978) found that it had about half the efficacy of ATP. Further studies suggested two (not mutually exclusive) reasons for this unexpected result. First, the medium used by Ishikawa et al. (1978) had been modified so that it did not maintain nuclear restriction quite so adequately (Agutter 1980). Second, when a brief (30 min) pre-labelling time is used, the ATP analogue has no effect on efflux; when a long (several hours) prelabelling time is used. $\beta\gamma$-methylene-ATP causes a significant stimulation of mRNA efflux (Jacobs and Birnie 1982). The efficacy of ATP itself is independent of the prelabelling time, though when longer prelabelling times are used, the time-course of RNA efflux shows significant efflux at zero minutes in the presence of ATP. These rather surprising observations proved to be explicable in ways consistent with the involvement of the NTPase, but they led to further insights into the translocation apparatus (see below).

3.   Protein Kinase and Protein Phosphohydrolase

As indicated above, it is well established that nuclear envelopes contain protein kinases (and phosphohydrolases) that regulate the phosphorylation of endogenous proteins. Studies by McDonald and Agutter (1980), aimed at explaining the surprizing effects of ATP analogues on mRNA efflux (see above), threw light on the possible involvement of these enzymes in translocation. Briefly, they found that poly(A) inhibits endogenous kinase activity and stimulates phosphohydrolase activity. The poly(A) concentrations involved are the same as those needed to stimulate the NTPase. This gave a prima facie case for arguing that kinase and phosphohydrolase activities were involved in translocation, and subsequent studies in other laboratories have supported this view (Bernd et al. 1982a; Goldfine et al. 1982a, b; Bachmann et al. 1984b). However, the relationships of these enzymes to the NTPase is presumably not direct, because the kinase activity, at least, shows a wholly different response to carcinogen feeding from the NTPase (Clawson et al. 1980c). It now appears that one phosphorylatable nuclear envelope polypeptide, molecular weight 110 kDa, shows a very marked sensitivity of its phosphorylation to poly(A) (Bachmann et al. 1984b; Agutter 1985c). In view of the work of Schweiger and his collegues on the 110 kDa poly(A) binding protein (see Sect. C.III.2) and of the evidence (below) for a single class of poly(A) binding sites in nuclear envelopes, this may be a particularly interesting finding. The finding by Bernd et al. (1982 a) that P78 prevents stimulation of the NTPase by poly(A) is also interesting. It might partly explain why mRNA transport is vectorial, given that P78 is restricted to the cytoplasm (see Sect. C.III.2).

4. The Poly(A) Binding Component

I remarked that the work of McDonald and Agutter (1980) was aimed at explaining why ATP analogues were effective efflux stimulators only when prelabelling times were long. What explanation did the work provide? Consider the following hypothesis.

Suppose that mRNA is displaced from its binding sites in the nuclear envelope, rapidly, by either ATP or a non-hydrolyzable analogue. However, occupation of these sites by mRNA is dependent on ATP hydrolysis. When a short prelabelling time is used, the sites are still occupied by unlabelled messengers when the nuclei are isolated. Thus, $\beta\gamma$-methylene-ATP does not cause efflux of label; ATP does, because it provides the energy for reoccupation of the sites by labelled material from within the nucleus. However, when a long prelabelling time is used, the sites are already saturated with labelled messenger. Under these conditions the non-hydrolyzable ATP analogue (and ATP itself) cause rapid displacement of labelled material into the medium, but only ATP itself can fuel the reoccupation of the sites.

McDonald and Agutter (1980) found that when labelled poly(A) was bound to isolated envelopes, both ATP and $\beta\gamma$-methylene-ATP displaced it. However, the poly(A) bound with higher affinity to phosphorylated than to unphosphorylated envelopes, and its binding promoted dephosphorylation. Overall, therefore, poly(A) binding seemed to require ATP, and poly(A) displacement did not. This gave general support to the hypothesis stated in the previous paragraph.

More importantly for an understanding of translocation, this work showed that mRNA [or poly(A)] binding was related to kinase and phosphohydrolase activities. Significantly, the dissociation constant for poly(A) binding to phosphorylated envelopes, derived from Scatchard plot analysis, was close to the poly(A) concentration needed for half-maximal stimulation of the NTPase in isolated envelopes. In view of this, and of corroborations from other laboratories (Goldfine et al. 1982; Bernd et al. 1982 a), it now seems that translocation of mRNA involves some kind of interplay between the NTPase, protein kinase and phosphohydrolase activities, and a mRNA binding site with some specificity for poly(A). This conclusion forms the basis of the kinetic models discussed in III (below).

5. The Nuclear Membranes

There is a widely held view that the nuclear membranes play no part at all in translocation, because removal of all the phospholipid (presumably with the intrinsic proteins) from the nuclear periphery does not affect mRNA efflux, and neither does lipid modification by a variety of in vivo or in vitro treatments (Clawson and Smuckler 1978; Murty et al. 1980; Palayoor et al. 1981; Agutter and Suckling 1982 a). However, two important caveats need to be entered. First, Yannarell and Awad (1982) obtained results that seem directly to conflict with those of Agutter and Suckling (1982 a). They found that greater efflux rates were obtained when the fatty acyls of the phospholipids were made more unsaturated, and Agutter and Suckling (1982 a) had found no such effect. Second, Smith and Wells (1984) obtained a soluble fraction from nuclear envelopes that contained the NTPase, and they found that in this form the enzyme depended for its activity on one minor component of the lipid — the phosphatidyl inositol (PI) and its derivatives. These two sets of observations merit discussion.

The main methodological difference between the studies of Yannarell and Awad (1982) and Agutter and Suckling (1982 a) was that the former group, but not the latter, included homologous cytosol proteins in the efflux media. These proteins include important stimulators of efflux (probably of translocation) which are, nevertheless, not mechanistically essential for translocation to occur (see Sect. E). Perhaps these stimulators affect translocation more efficiently when the lipids are unsaturated. If so, it is possible that the detailed chemistry and the physical state of at least some of the nuclear envelope lipid is important for regulating translocation, and hence the efflux and presumably the transport of mRNA. This is generally consistent with the findings of Smith and Wells (1984). PI and its derivatives play crucial roles in metabolic regulation, and evidence suggesting that they regulate mRNA translocation is not inherently surprising. A possible implication of these two publications, therefore, is that cytoplasmic regulators of translocation act inter alia by modulating PI metabolism in the nuclear envelope. This issue will be discussed further in III, below, and Section E.

Granted these important caveats, it remains established that the bulk of the nuclear envelope lipid (and probably most of the protein components of the membranes) play no part in translocation, at least in higher eukaryotes.

6. The Pore-Complexes

The most important single barrier to progress in our understanding of translocation is the ill-characterized state of the pore-complexes. Numerous attempts have been made to obtain satisfactory isolated pore-complex preparations and to identify their main protein components which can then be labelled by specific antibodies. Except for the unequivocal identification of the 190 kDa glycoprotein (Sect. B.I.), all these attempts have failed. This state of affairs is particularly frustrating because the prob-

lem looks fairly straightforward given modern biochemical techniques. Appearances
are deceptive. The problem continues to baffle all workers in the field. It is clear that
a fully adequate account of translocation will not be obtained until this problem is
solved.

Two approaches that should perhaps be tried are the following. First, the possibility
(a likely one on the face of it) that the poly(A) binding component is part of the pore-
complex could lead to some methodological advances, based on a covalent cross-linking of poly(A) to the binding protein and subsequent affinity chromatography. The
low-affinity colchicine-binding component might also behave as a usable pore-complex marker (Agutter and Suckling 1982b). Second, if the phosphorylation-depen-
dent mitotic breakdown of the nuclear envelope could be simulated in vitro, conditions might be established that would liberate the pore-complexes without disrupting
them. The factors that are known or suspected to be important in this breakdown process are lamin B phosphorylation, as described by Gerace and his colleagues (see
Sect. B.I), and removal of the 33 kDa polypeptide that has been called periplasmin
from the envelope (McKeon et al. 1984). Perhaps deployment of these and other approaches will bring the problem of pore-complex characterization nearer to comple-
tion in the next year or two.

III. Towards a Model for Translocation

1. The Evolution of Kinetic Schemes

When several components are known to interact in a biochemical process, it can be
helpful to advance a hypothesis about their mode of interaction in the form of a kinetic scheme. Such a scheme, if appropriately conceived, leads to critically testable predictions. When the scheme is tested, it is usually refuted or modified in the light of results, and this constitutes advancement of knowledge. Clearly, many components interact in mRNA translocation, and kinetic models are therefore likely to play an important part in the growth of our understanding.

If kinetic schemes serve their purposes best by being refuted, then the scheme proposed by Agutter (1980) and McDonald and Agutter (1980) was philosophically excellent: it was refuted rapidly and comprehensively. This scheme, advanced to explain
the ATP-analogue results (Ishikawa et al. 1978), suggested that the NTPase did not
exist independently, but simply represented the combined activities of the protein
kinase and phosphohydrolase. ATP binding displaced the bound RNA, and subsequent phosphorylation increased the affinity of the system for more mRNA from
the nuclear compartment. Refutation is simple: the maximum catalytic rates of the
NTPase and the protein kinase differ by an order of about 1000 (Bernd et al. 1982a).

Subsequent schemes, however, were explicitly advanced as modifications of the
Agutter (1980) scheme. Goldfine et al. (1982) proposed that the NTPase was directly
inhibited by protein kinase-dependent phosphorylation and stimulated by dephos-

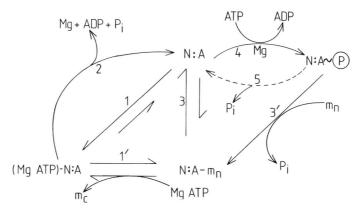

Fig. 8. The most recent kinetic scheme of translocation (Agutter 1985c). *N NTPase; A* binding site for mRNA [poly(A)]; m_n, m_c messenger in nuclear and cytoplasmic compartments, respectively. Reactions *1, 1'* substrate binding to the NTPase active site; displacement of mRNA at the cytoplasmic surface is concomitant with this (1'). Reaction *2* ATP hydrolysis by the NTPase. Reactions *3, 3'* messenger binding at the nuclear face of the system. The affinity for the phosphorylated binding site (*3'*) is greater than that for the unphosphorylated site (*3*), and binding to the phosphorylated site is accompanied by dephosphorylation. Reaction *4* phosphorylation of the binding site by the protein kinase. Reaction *5* phosphohydrolase activity. This activity is low unless stimulated by mRNA binding (*3'*)

phorylation. This agreed with most of the known facts, and accommodated the concept of insulin stimulation of the NTPase (via stimulation of the phosphohydrolase), but it did not account for the effect of phosphorylation on mRNA [or poly(A)] binding affinity. Later, Bachmann et al. (1984b) proposed a more complex scheme, allowing for poly(A) binding effects. This scheme, however, implied that phosphorylation stimulated rather than inhibited the NTPase and hence translocation (see also Mueller et al. 1985).

Agutter (1985c) has proposed the most recent kinetic scheme (a version is shown in Fig. 8). This seeks to combine the best-corroborated features of the other schemes advanced so far, and takes into account the important finding that the NTPase is not itself a protein kinase substrate. This finding, which was based on photoaffinity labelling studies with 8-azido-ATP and phosphorylation studies with phosphate-labelled ATP, was made almost simultaneously in two laboratories (Clawson et al. 1984; Agutter 1985c). According to these papers, the NTPase is a protein of 43–47 kDa. This clearly distinguishes it from the 174 kDa ATPase/dATPase (Berrios et al. 1983).

Amongst the testable predictions of this latest scheme are:

a) The purified isolated NTPase will be insensitive to both phosphorylation and poly(A).

b) The purified enzyme will display linear Michaelis-Menten kinetics under assay conditions in which it deviates from such kinetics in whole nuclear envelopes.

A most important step in our recent progress in this field has been the purification of the NTPase to homogeneity by Mueller's group (Schroeder et al. 1986). So far, these two predictions of the scheme have been corroborated by the findings of this group. Moreover, their molecular weight estimate agrees with those from the photoaffinity labelling studies.

Two other features of the scheme in Fig. 8 are as follows. First, it allows for "short circuits", in which either the phosphorylation-dephosphorylation cycle or the NTPase reaction itself play no part in translocation. These are relevant in some cancer cells, where mRNA efflux is ATP-independent (Schumm et al. 1977; Otegui and Patterson 1981), but it is not yet possible to explain how the transition to ATP-independence comes about (i. e. how and why the short circuiting occurs). Second, it allows for control of translocation, as all schemes subsequent to that of Agutter (1980) have done. Specifically, increse of the protein kinase activity will (according to the scheme) inhibit the NTPase in the absence of messengers [or poly(A)], but stimulate it in the presence of these materials. The possibility of control of translocation via the protein kinase as well as the phosphohydrolase will be discussed further in Section E, but it can be noted that protein kinase stimulation dependent on the PI cycle, and perhaps on cytoplasmic stimulators of mRNA efflux, can be accommodated in the scheme (cf. D.II.5, above).

2. Advantages and Limitations of Kinetic Schemes

Some of the advantages of schemes of this kind were outlined in (1), above. Provided that they are not taken as literal descriptions of reality, they are at worst harmless. Normally, they are useful. The evolution of kinetic schemes of translocation shows how they have served to unify existing information, predict new findings and thus suggest further experiments, and lend themselves to modification (or outright rejection) in the light of such experiments. From the biochemist's point of view, they are well-nigh indispensible for these reasons. It can also be noted that from empirical values of affinities and rate constants, the predictions of such schemes can be made quantitative and lend themselves to computer simulation of specific experiments. At this point, it becomes fair to call them models rather than merely schemes. (I am currently using the scheme in Fig. 8 in precisely this way).

However, this approach to the field of mRNA transport has limitations. First, the schemes discussed here pertain to only one part of transport — translocation — because we know too little about the other stages of transport even to begin to model them. Second, kinetic schemes (or models) do not contain any topological information. Because mRNA transport is vectorial (and translocation itself is apparently vectorial, according to the resealed-envelope studies of Riedel et al. (1987)), and because it involves objects with precise geometries (the nuclear envelope, and specifically the pore-complexes), topological information is an essential ingredient in our understanding. Kinetic schemes or models can never supply this ingredient, and for this reason, if for no other, it is a mistake to rely too heavily upon them. However, in our present state of progress the poorly characterized state of the pore-complexes (Sect. D.II.6;

see also Sect. B.I) means that we simply do not have the topological or geometrical information in a form that we can use for analyzing translocation at the biochemical level. Therefore, kinetic schemes have taken on a greater appearance of value than they might otherwise have had.

Another point that Fig. 8 does not take into account is the involvement of the lamins, especially lamin B. If the antibody used by Baglia and Maul (1983) was indeed against lamin B, not against another protein of identical molecular weight, then this major nuclear envelope protein component is clearly relevant to translocation. In this context, the likelihood that the NTPase is located in the inner nuclear membrane along with the PI (Smith and Wells, 1984; M. Schindler, pers. commun.) may be very important. Lamin B is so intimately associated with the inner membrane that it is very likely to be closely coupled to the NTPase, and (perhaps along with the rest of the lamina) it might form a mechanical coupling system linking the energy-transduction site (the NTPase) to the mRNA binding and translocating site (presumably the pore-complex, from which the lamina is inseparable by current isolation techniques). The finding that oxidative cross-linking of proteins impairs translocation (Zbarsky and Peskin 1982; Kindas-Muegge and Sauermann 1985) might provide corroborating evidence for this view, because the lamins are known to be susceptible to such cross-linking (Kaufmann et al. 1983). These arguments, unlike those from kinetic schemes, lead us closer towards a topological description of the translocation apparatus. It seems likely that in vitro reconstitution studies using purified NTPase, lamin B and poly(A) binding protein (minimally) will provide more definitive information in due course. This line of approach is being considered at present in more than one laboratory.

E. The Control of Messenger Transport

Webb et al. (1981), Lichtenstein et al. (1982) and Agutter (1984) have reviewed the control of mRNA transport in detail, and for the purpose of this chapter it seems more appropriate to try to relate control to what we know about the mechanisms of transport. Broadly, mRNA transport is one step in gene expression and protein biosynthesis, so I shall first try to establish its significance in this general context. Second, it seems likely (see Sect. D.I.1) that a good deal of control of mRNA transport is located at the release stage. This is problematic for a reviewer, because in our present state of knowledge it is not possible to distinguish release-specific control from the control of post-transcriptional HnRNA processing (there may in fact be no difference). In any case, we know far more about translocation and what possibilities it offers for control. This state of affairs determines the remainder of what I shall say in this section.

I. The Control of Gene Expression

1. Overview

Our present understanding suggests that gene expression is controlled at five distinct levels:

a) Transcription.
b) Post-transcriptional HnRNA processing.
c) mRNA transport.
d) Translation.
e) Stability and turnover of cytoplasmic messenger pools.

   Obviously, I am not including post-translational events and protein modifications, or gene rearrangements, that are themselves enormously important in the control of metabolism. The five levels I have noted could, potentially, play significant parts in the control of cellular differentiation, a point that is discussed by Lichtenstein et al. (1982) and Agutter and Thomson (1984). It would be pointless to try to offer anything like a comprehensive review of the literature pertaining to any of these five levels in the space available here, let alone an extensive discussion of all of them. Instead, I shall briefly indicate what is known about the relative significance of each level.

   a) Transcription is crucially important. This well-known fact is emphasized, for instance, in reviews by Paul (1982) and Darnell (1982), the latter of which perhaps places too little emphasis on our knowledge of other control levels. Transcription can be seen as a "coarse control" of gene expression and cell differentiation; the other four levels participate in "fine tuning" of cellular responses. A point that is interesting in the context of the present review is that cytoplasmic factors, apparently proteins, regulate transcription, and therefore nucleocytoplasmic exchange mechanisms are relevant to this most fundamental aspect of the control of gene expression. The evidence for this was indicated in the last part of Section B.II., and is discussed in detail by Lichtenstein et al. (1982).

   b) General evidence that post-transcriptional processing and/or mRNA transport are significant control steps in gene expression is provided by the differences in sequence complexities between HnRNA and mRNA, though it should be remembered that differential mRNA stabilities could account for some of these data. However, not only is transcription susceptible to different cytoplasmic regulators and different inhibitors from processing and transport events (McNamara et al. 1975; Hazan and McCauley 1976; Weck and Johnson 1976, 1978; Gajdardjeva et al. 1980), but there is a wealth of evidence for control levels between transcription and translation from studies of morphogenesis, tissue regeneration, virus infection, neoplastic transformation, and hormonal control (see Lichtenstein et al. 1982; Agutter 1984 for detailed reviews). Finally, the cell's repertoire of proteins can be increased by splicing single gene products differently (see citations in Sect. C.I.3), and this clearly indicates the actuality of control of processing per se.

c) Although much of the available evidence is ambiguous in that control could be exerted either at the processing or the transport stages, some studies very clearly indicate the reality of transport control per se. Hereditary hypothalamic goitre in goats is characterized by normal transcription and processing of an apparently normal thyroglobulin gene, but the mature 33S messenger accumulates in the nuclei and its cytoplasmic levels are 20 to 50 times lower than normal (Van Voorthuizen et al. 1978). There is evidence that insulin modulates the transport of some high-abundance messengers in liver (Purrello et al. 1982), and that 3',5'-cyclic AMP activates at least one of the cytosol-derived liver proteins that specifically enhances mRNA transport, not HnRNA processing (Schumm and Webb 1978; Moffett and Webb 1983). These points are discussed further in (2), below, and in more detail in Subsection II.

d) It is well established that translation is an important level of control (see e.g. Darnell 1982). Generally, it seems that the formation of initiation complexes is a regulatable event, and that messengers can, in response to hormonal or other appropriate signals, move from the non-translatable to the polysomal pool or vice-versa. Given the central relevance of cytoskeletal binding to translation, it seems that control of the final stage of mRNA transport is perhaps indistinguishable from the control of translation, just as control of the first stage (release) is indistinguishable from processing.

e) Over the past few years, it has come to be generally held that control of mRNA stability and degradation is second in importance only to transcription as a means of regulating gene expression. Some recent contributions to this belief include the destabilization of type I procollagen mRNA in skin fibroblasts by cortisol (Hämäläinen et al. 1985), stabilization of amylase messenger in pancreatic acinar cells by glucocorticoids (Logsdon et al. 1985), and the destabilization of non-globin messengers during erythroid transformation (Krowczynska et al. 1985). These exemplify a steadily growing literature. Particularly interesting in the context of this chapter is the finding by Khalili and Weinmann (1984) that adenovirus infection has two opposite, mutually compensating effects on actin gene expression in HeLa cells. The stability of the actin messenger in the cytoplasm is greatly enhanced by the virus, but nucleocytoplasmic transport is greatly inhibited. The overall effect is approximately zero. Transcription itself is not affected. The implication of this study is that the control of mRNA stability is mechanistically inseparable from nucleocytoplasmic transport. This recalls the image of an intracellular space that is structurally and functionally integrated by systems of fibrils (see Sect. C.III and the introduction to Sect. D). Such an image is obviously compatible with the solid-state perspective which I believe to be inescapable in the study of mRNA transport.

2. The Place of mRNA Transport

In the very brief overview I have just given, I have suggested that (a) mRNA transport per se is important in some systems in regulating gene expression, (b) aspects of mRNA transport control may be indistinguishable in principle (as, at present, they certainly are in practice) from control of processing, translation, and even mRNA stability, (c) other nucleocytoplasmic exchanges, particularly protein exchanges, are important in regulating intranuclear events, including translation itself. It is clear that there are problems in proposing sharp lines of demarcation between the various levels of control of gene expression. The more evidence becomes available about the interactions of the five levels, the more the cell appears as the kind of integrated entity that lends itself to a solid-state description.

If we restrict our attention to those cases where mRNA transport is controlled but post-transcriptional processing and translation are apparently not, then we are left with a relatively small literature from which it appears that mRNA transport is only of minor significance overall. I shall restrict my attention in Section D.II, below, mainly to such cases, viz: insulin and 3',5'-cyclic AMP dependent controls in liver, steroid-dependent control in oviduct, impaired thyroglobulin messenger transport in thyroid cells, specific cytoplasmic protein factors and their alterations incident on carcinogenesis and other pathological events, and age-related changes. I shall try to establish which of these involve events before translocation, and which can be related to the translocation mechanism itself. I shall discuss the latter with reference to our understanding of translocation (Sects. D.II and D.III).

II. The Loci of Transport Control

1. Events Before Translocation

I have already indicated that the control of processing (notably splicing, which seems to be the rate-determining step: Mariman et al. 1982) might be difficult or impossible to distinguish in many cases from the control of the release stage of transport. (The same confusion applies to translation and cytoskeletal binding). Most of the studies that have definitely shown control of transport, not of processing or translation, do not suffer from this problem: they show effects on a specific component of the *translocation* system, typically the NTPase. However, the work of Van Voorthuizen et al. (1978) raises interesting questions that are not yet answered. Specifically, if the 33S messenger that accumulates in the nucleus is really fully mature, does it lack an appropriate "permeation" signal and therefore the ability to engage functionally with the translocation apparatus? Or has it failed to be released from the intranuclear reticulum? If the former, then the "permeation" signal in this case must be subtler than the mere presence of a poly(A) tail, and we have no idea what it could be. If the latter, then there must be a nuclear "accumulation" signal other than introns. In any case, the release and translocation mechanisms cannot be inherently faulty, because transport of other messengers in these cells appears to be quite normal.

The nub of the issue with this work is that the state and location of the 33S messenger in the nucleus is not known. No circumstance could better illustrate the extent of our ignorance about the events responsible for release. We can hope that when the intranuclear reticulum is better characterized, we shall at least be able to make some concrete suggestions about the reasons for this kind of pathological nuclear restriction. Meanwhile, the etiology of hypothyroid goat goitre itself deserves further study at the molecular level.

As a more general illustration of the problem of locating regulatory events precisely, the studies of post-transcriptional control of gene expression in neurons and neurotumors by Beckmann et al. (1981) or in duck erythroblasts (Imaizi-Scherrer et al. 1982) can be considered. These groups conducted extensive hybridization studies to characterize nuclear and cytoplasmic RNA. The work is undoubtedly thorough. The conclusion is that only about 30 % of transcribed sequences are eventually translated in the brain system, and about 5 % in the erythroblasts, indicating considerable post-transcriptional selection. Differential stabilities of cytoplasmic messengers do not explain these results, but the authors were unable to offer any firm suggestion about the respective roles of processing, transport and translation. Whichever level is important (all three might be), one could, for the reasons outlined in Section I.1, above, argue that the major control step in this system is transport-related. Having said this, we would be no closer than we were before to a satisfactory understanding.

Benyumov et al. (1983) attempted to characterize the role of post-transcriptional control events by experiments in which embryonic loach nuclei were transplanted into unfertilized eggs. They concluded that abnormal development of the embryos resulting from nuclear transplantations can be attributed to disturbances of mRNA transport. However, their results could equally well be interpreted in terms of impaired processing (and release) or impaired translation (and cytoskeletal binding). Once again, it is not possible to identify the locus of control precisely enough to make the conclusions useful. This study is especially worth considering by readers who are disposed to think that in situ studies might elucidate the issues more satisfactorily than in vitro studies.

2. Translocation

Studies that definitely show control at the level of translocation lead to a discussion that contrasts happily with the rather pessimistic content of the previous paragraphs. The literature in this area can be classified into (a) studies that show effects on the components of the translocation system in the nuclear envelope itself and (b) studies that show effects mediated by cytosolic regulatory proteins. There is continuing progress in both these subdivisions of the topic. A review of them shows that substantive conclusions can, after all, be drawn about the control of gene expression at the mRNA transport level.

a) Control Mediated Through Nuclear Envelope Components

Insulin stimulates mRNA efflux from isolated liver nuclei, just as it increases the rates of accumulation of some high-abundance messengers, such as albumin mRNA, in vivo (Schumm and Webb 1981). The levels of insulin that stimulate efflux, which are close to physiological levels, also stimulate the NTPase in isolated nuclear envelopes, and they seem to do this by stimulating the protein phosphohydrolase activity (Purrello et al. 1982; Goldfine et al. 1982; Purrello et al. 1983). The scheme in Fig. 8 suggests that enhanced dephosphorylation will increase the NTPase activity but decrease the affinity of the binding site for mRNA, without altering the total translocation capacity. A decreased affinity could mean that the transport of low-abundance messengers is retarded (their availabilities are low compared to the increased dissociation constant) and therefore more of the total translocation capacity becomes available for high-abundance messengers, such as albumin mRNA. In this way, the results seem to be compatible with our understanding of translocation. Other studies that have shown hormonal modulations of the NTPase have come from Mueller's group. Bernd et al. (1983) showed that oestrogen and progesterone increased the NTPase activity about eight fold in immature quail oviduct nuclear envelopes, but had no effect in liver. The increase correlates with a huge increase in cell proliferation and of ovalbumin and avidin synthesis. Similar changes were found during normal maturation, and the reverse changes (correlating with decreased hormone levels) occurred as the animals aged. This group has also reported a 50% increase in NTPase activity in skin fibroblasts in response to dexamethasone, correlating again with changes in protein biosynthesis in vivo (Bernd et al. 1984; cf. also Kaufmann and Shaper 1984). It is not yet clear whether these steroid hormones interact directly with the NTPase or with some other part of the translocation apparatus.

The NTPase activity itself decreases with increasing age of the organism, but more interestingly the number of poly(A) or mRNA binding sites in the nuclear envelope decreases, and the capacity of poly(A) or mRNA to stimulate the NTPase is drastically impaired, with increasing age (Bernd et al. 1982b). The actual affinity of the binding-site for poly(A), and its phosphorylation dependence, remain unchanged between young/mature and old animals. The loss of mRNA binding capacity might parallel the decrease in numbers of pore-complexes per unit area. The loss of functional coupling to the NTPase is less easily explained, but it would be interesting to look for changes (e.g. endogenous cross-linking) in the lamina (cf. final paragraph in Sect. D). Alternatively, changes in the nature or efficacy of the protein kinase could usefully be investigated.

Although their relevance (if any) to the translocation apparatus has not been established, a variety of protein kinases have been and continue to be reported in the literature. Mednieks and Hand (1983) found a 3',5'-cyclic AMP-dependent kinase in rat parotid acinar nuclei, and although its precise location was not established, association with the pore-complex or lamina was not ruled out. A kinase that is insensitive to cyclic nucleotides was reported in nuclear envelopes by Iyer and Mastro (1982). The kinase that is involved in the translocation scheme in Fig. 8 is not responsive to cyclic nucleotides, but Clawson et al. (1980) reported that the NTPase itself is stimulated by 3',5'-cyclic AMP. This could mean that more than one kinase is, after all, relevant, and the kinase that is sensitive to cyclic nucleotides exerts a stimulatory rather than an inhibitory effect on the NTPase. As yet, this view is uncorroborated, and the scheme in Fig. 8 has not been further complicated by incorporating it.

This list of findings is illustrative, not exhaustive. Possible future developments include further insights into the regulatory significance (if any) of the effects of heat-shock proteins on nuclear envelopes (Smith and Fisher 1984).

b) Control Mediated via Cytosolic Protein Factors

I raised the issue of protein kinases again in the previous paragraph partly because they could be relevant to translocational control by cytosol protein factors. Webb and his colleagues have conducted a long series of studies on these factors, and their earlier experiments were reviewed by Webb et al. (1981) and Agutter (1984). Recent progress, mainly reflected in publications from Webb's group, has included purification of the major factors from normal and carcinogen-treated rat livers and other tissues. In other laboratories, there have been insights into the possible mode of action of the proteins concerned. As yet, we do not know whether "cytosol" implies that the proteins are cytoplasmic in vivo: labelling with specific antibodies has not been described.

Moffett and Webb (1981, 1983) showed that the major mRNA transport factor from rat liver could be obtained from polysomal preparations and purified by differential salt extraction and column chromatography. These results have been corroborated in my laboratory: McDonald (1984) purified the major factor from normal and carcinogen-transformed livers by techniques that differed only in detail from those of Moffett and Webb (1983). The major factor is a 30–35 kDa acidic phosphoprotein, highly thermolabile and present in very low concentrations in cytosol or polysomal preparations. It is not highly conserved: Smart-Nixon et al. (1983) showed that it had considerable species and even tissue specificity.

Carcinogen treatment alters the concentration, complexity and form of the cytoplasmic mRNA population (see e.g. Shearer 1974, 1977; Clawson and Smuckler 1982 a, b). This suggests an altered regulation of, among other things, mRNA transport. Webb and his colleagues have duly reported that a 60 kDa protein factor appears in cytosol preparations after carcinogen feeding. This protein seems to be responsible for the altered mRNA efflux from nuclei in vitro, which reflects carcinogen-induced changes in vivo (Walaszek et al. 1983; French et al. 1984). It is not clear whether this 60 kDa protein is specifically associated with carcinogenic alterations of mRNA transport or whether an incidental effect of a stress protein has been observed. McDonald (1984) found that 3,4-dimethylaminoazobenzene-induced liver transformations were associated with an increased quantity or activity of the 30–35 kDa factor, but he found no evidence for other specific factors. Mueller and his colleagues have found a 54 kDa

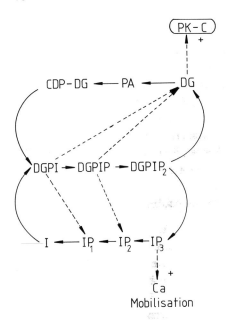

Fig. 9. A version of the phosphatidyl inositol (PI) cycle, based on views developed in the early 1980's *DG* diacylglycerol; *PA* phosphatidic acid; *DGPI* phosphatidyl inositol. Further phosphorylation of the inositol moiety generates the other forms shown, all of which can be hydrolysed by phospholipase C to diacylglycerol and the corresponding inositol uni-, bis-, or tris-phosphate. The tris-phosphate is believed to be the agent that mobilizes calcium stores; the diacylglycerol activates protein kinase C, which is itself calcium-dependent. These two activation pathways, calcium mobilization and protein phosphorylation, can act independently or synergistically. This scheme is based on views established in 1984, but with the discovery of still higher phosphates of inositol it might be obsolescent

protein apparently capable of stimulating mRNA efflux, as well as the 30–35 kDa factor, in normal livers (W.E.G. Mueller, pers. commun.). No other laboratory has reported this.

Modifications of the main 30–35 kDa protein seem to be responsible for several types of control of translocation. For instance, it is activated by 3',5'-cyclic AMP-dependent phosphorylation (Moffett and Webb 1983; cf. Schumm and Webb 1978), and this could explain the stimulation of efflux by the cyclic nucleotide. (Moffett and Webb (1983) made the rather odd suggestion that this cytosol factor phosphorylation could explain the ATP dependence of mRNA efflux, overlooking — among several other things — the relative quantities of ATP involved). Chronic ethanol poisoning leads to a partial loss of nuclear RNP restriction in rats, apparently (in part) because of alterations in the cytosol factor. These points make the mode of action of the factor well worth studying at the biochemical level.

McDonald (1984) found that the 30–35 kDa factor increased the affinity of the poly(A) binding site in phosphorylated, but not unphosphorylated, nuclear envelopes, and correspondingly facilitated the poly(A) stimulation of the NTPase. Its ability to increase poly(A) affinity was enhanced by 3',5'-cyclic AMP-dependent phosphorylation. This result makes two inferences possible. First, it confirms that the cytosol factors do indeed regulate translocation. (In this context, it is interesting that the 54 kDa factor apparently stimulates the NTPase in the absence of poly(A): W.E.G. Mueller, pers. commun.). Second, it suggests that what is primarily stimulated is the protein kinase. How this happens has not been demonstrated, but one rather exciting possibility is that the factor stimulates the PI cycle in the inner nuclear membrane (Fig. 9). This would be consistent with the suggestion by Smith and Wells

(1984) that PI cycle-related regulation of the NTPase could explain, among other things, the well-known changes in this enzyme incident on carcinogenesis (see Sect. D.II.1). If this interpretation is correct, then the kinase is presumably protein kinase C, and we would expect the other arm of PI-related control, i.e. mobilization of calcium, to be in evidence in the same system. (For current views of the PI cycle, see Berridge 1984; Berridge and Irvine 1984; Nishizuka 1984). These predictions are testable, and are currently being studied.

Whatever the outcome of such studies, one thing is clear: information about the *control* of mRNA transport can, when analyzed biochemically, elucidate transport *mechanisms,* just as knowledge of the mechanism makes the possible loci of control clearer. Perhaps we can hope for similar reciprocal advances with regard to stages of mRNA transport other than translocation within the next decade.

F. Conclusions

1. General Summary

In this chapter, I have tried to introduce the reader to the current status of our understanding of nucleocytoplasmic mRNA transport and to delineate current and possible future progress in this field. I have tried to relate studies on the mechanism and control of mRNA transport per se to our understanding of and progress in other relevant fields: nucleocytoplasmic exchanges of other macromolecules, HnRNA processing and mRNA metabolism, RNP structure, intracellular (and especially intranuclear) skeletal structures, nuclear envelope ultrastructure and biochemistry, and the control of gene expression and cellular differentiation. I believe this broad, multidisciplinary range to be essential for a coherent understanding of the field, its successes and failures, and its current directions and difficulties.

The main points that I have tried to make about mRNA transport are, briefly, these.

a) It is a solid-state process, in which the transported material is never diffusible in any physiologically, mechanistically important sense. This contrasts with other nucleocytoplasmic exchange processes, but some studies of these other processes have provided concepts that are relevant to mRNA transport, notably the significance of "association" and "permeation" signals.

b) It forms part of a continuum of biochemical events that extends from HnRNA processing to translation and messenger degradation. Thus, early events in mRNA transport are epistemologically, and probably mechanistically, inseparable from processing. To understand transport, it is therefore necessary to understand processing and HnRNP structure and organization. Similarly, late events in transport are inseparable from translation and from the dynamics of the cytoskeleton. Studies of these processes are therefore also relevant to mRNA transport.

c) A solid-state process requires a solid-state framework. Thus, issues of characterization not only of the cytoskeleton and the pore-complexes, but also the putative "nuclear matrix", become relevant to mRNA transport. The ill-characterized status of both the pore-complex and the "matrix" is currently a serious barrier to progress, and as yet the nature of mRNP-cytoskeleton linkage has not been adequately analyzed. These are obvious subjects for future research.

d) The best-understood stage in mRNA transport is the stage that many workers have taken to be the whole of transport, i.e. translocation across the nuclear envelope. The following points can be taken as established:

1. mRNA translocation occurs through the pore-complexes.
2. In most systems it is energy-requiring. The key energy transducer is the nucleoside triphosphatase (NTPase), an approximately 45 kDa protein probably located in the inner nuclear membrane, and recently purified from rat liver. Some cancer cells are exceptional in that mRNA translocation in them is ATP-independent.
3. The NTPase is modulated by protein kinase and phosphohydrolase activities intrinsic to the envelope. The kinase might be protein kinase C, and its activation related to phosphatidyl inositol cycle intermediates.
4. These activities also modulate the affinity of a unique class of RNA binding sites in the envelope that have some specificity for poly(A). These sites might be provided by a 110 kDa protein, the phosphorylation of which is highly poly(A)-sensitive.
5. One of the main components of the perinuclear lamina, lamin B, appears to be intimately associated with the NTPase. It might couple this enzyme to the mRNA [poly(A)] binding site.

e) Much of the control of mRNA transport involves the translocation apparatus, which responds specifically to hormones, carcinogens and other xenobiotics, and ageing. Some of this control is exerted via protein factors that are recovered from soluble or polysomal preparations from tissue homogenates. Their intracellular location is not known.

f) Messenger transport appears to play a rather minor part in general in the control of gene expression. However, given (1) the solid-state perspective and (2) the lacunae in our present understanding, clear distinctions between messenger transport and other, undoubtedly important, levels of control cannot be made.

The inadequacy of our present understanding needs to be emphasized. It is far from clear why post-transcriptional processing has to be completed (usually) before mRNA appears in the cytoplasm. When HnRNA-associated proteins are replaced by messenger-associated proteins, let alone how (and why) such replacement occurs, are mysteries. The molecular basis for the non-diffusibility of HnRNP and mRNP is unknown. The pore complex remains, biochemically speaking, a question mark. It will be a long time before these large areas of ignorance are resolved.

Nevertheless, recent progress has been useful and promising. The translocation system is now fairly well understood, and will probably be still better understood within a very few years. The prolonged controversy about the nuclear matrix now appears to have transmuted, at least, into a clearly defined problem: the characterization of the dense intranuclear reticulum. The problem remains difficult, but at least it is articulated. At least some of the "cytosol" factors have been purified, and understanding of their mechanisms of action have advanced greatly over the past 2 or 3 years. Meanwhile, advances in the general study of nucleocytoplasmic transport, HnRNA processing and RNP strucutres have illuminated the periphery of the field and clarified the general context of mRNA transport studies.

2. Some Contentious Points

It is important that newcomers to a field should be aware of a reviewer's biases and idiosyncracies so that they can assess the review article in a properly critical way. Apart from the speculations and the ideas for future progress that I have offered, I am aware of four particularly tendentious claims in this article:

a) Messenger transport is claimed to be a solid-state process. For this reason, diffusion-related concepts such as pore-complex permeability are irrelevant to it. Not everyone would agree.

b) Poly(A) is claimed, controversially, to have a role in mRNA transport, (though it is not a sufficient or even a uniquely necessary condition for transport).

c) It is claimed that some kind of "nuclear matrix" exists in situ, though many isolated preparations are more or less artefactual, and that this "matrix", or nuclear reticulum, is centrally relevant to mRNA transport. This is another highly controversial claim.

d) It is claimed that in vitro methods for studying mRNA efflux are, within certain defined limits, not only well-characterized and acceptable models for mRNA transport, but indispensible for studying the process. There are still workers who question this. Conversely, in situ methods are incapable of elucidating mRNA transport usefully, despite their considerable successes in other areas.

I believe that these four claims are well justified by the evidence I have discussed at appropriate points in this article. However, workers who dissent from these claims think otherwise, and the reader should study their reasoning carefully before deciding whether such claims should be accepted. Critical appraisal of this kind is the essence of scientific progress, and the onus is on exponents of particular positions to be explicit about the claims they are making and why they are making them. The resulting debate and conceptual advancement are as important to the progress of controversial fields, such as mRNA transport, as the design of novel methods and the execution of well-conceived and informative experiments.

Acknowledgement. I am very grateful to Mr. W. Lewicki and to Dr. A.M. Tunstall for preparing the diagrams, to Mrs. S.A. Comerford and Mrs. G. Ollerhead for supplying the electron micrographs in Fig. 6, and to Dr. A.M. Tunstall for her help in preparing the manuscript.

References

Aaronson RP, Blobel G (1975) Isolation of nuclear pore complexes in association with a lamina. Proc Natl Acad Sci USA 72:1007–1011

Aaronson RP, Coruzzi L, Maul GG (ed), Schmidt K (1982) Nuclear pore complexes: status of isolation attempts. In: The nuclear envelope and the nuclear matrix. Alan R Liss, New York, pp 13–20

Acheson NH (1984) Kinetics and efficiency of polyadenylation of late polyoma virus nuclear RNA: generation of oligomeric polyadenylated RNA species and their processing into messenger RNA. Mol Cell Biol 4:722–729

Adeniyi-Jones S, Romeo PH, Zasloff M (1984) Generation of long read-through transcripts in vivo and in vitro by deletion of 3' termination and processing sequences in the human initiator tRNA-met gene. Nucl Acids Res 12:1101–1115

Adesnik M, Salditt M, Thomas W, Darnell JE (1972) Evidence that all mRNAs molecules (except histone mRNA) contain poly(A) sequences and that the poly(A) has a nuclear function. J Mol Biol 71:21–30

Adolph KW, Kreisman LR (1983) Surface structure of isolated metaphase chromosomes. Exp Cell Res 147:155–166

Adolph KW, Cheng SM, Laemmli UK (1977) Role of non-histone proteins in metaphase chromosome structure. Cell 12:805–816

Agutter PS (1980) Influence of nucleotides, cations and NTPase inhibitors on the efflux of ribonucleic acid from isolated rat liver nuclei. Biochem J 188:91–97

Agutter PS (1982) Comparison of methods for studying RNA efflux from isolated nuclei. In: Maul GG (ed) The nuclear envelope and the nuclear matrix Alan R Liss, New York, pp 91–110

Agutter PS (1983) An assessment of some methodological criticisms of studies of RNA efflux from isolated nuclei. Biochem J 214:915–921

Agutter PS (1984) Nucleocytoplasmic RNA transport. Subcell Biochem 10:281–357

Agutter PS (1985a) RNA processing, RNA transport and nuclear structure. In: Clawson GA, Smuckler EA (eds) The nuclear envelope and RNA maturation UCLA Symposium, vol 26 (in press) Alan R Liss, New York, pp 539–560

Agutter PS (1985b, 1986) Isolation and subfractionation of nuclear envelopes. In: Birnie GD, Macgillivray AS (eds) Nuclear structures; their isolation and characterisation. Butterworth, London, pp 34–46

Agutter PS (1985c) The nuclear envelope NTPase and RNA translocation. In: Clawson GA, Smuckler EA (eds) UCLA Symposium vol 26. Alan R Liss, New York, pp 561–578

Agutter PS, Birchall K (1979) Functional differences between mammalian nuclear matrix and pore-lamin preparations. Exp Cell Res 124:453–460

Agutter PS, McCaldin B (1979) Inhibition of RNA efflux from isolated SV40–3T3 cell nuclei by 3'-deoxyadenosine (cordycepin). Biochem J 180:371–378

Agutter PS, Ramsay I (1979) Further studies on the stimulation of the nuclear envelope NTPase by polyguanylic acid. Biochem Soc Trans 7:720–721

Agutter PS, Richardson JCW (1980) Nuclear non-chromatin proteinaceous structures: their role in the organization and function of the interphase nucleus. J Cell Sci 44:395–435

Agutter PS, Suckling KE (1982a) The fluidity of the nuclear envelope lipid does not affect the rate of nucleocytoplasmic RNA transport in mammalian liver. Biochim Biophys Acta 696:308–314

Agutter PS, Suckling KE (1982b) Effect of colchicine on mammalian liver nuclear envelope and on nucleocytoplasmic RNA transport. Biochim Biophys Acta 698:223–229

Agutter PS, Thomson I (1984) Nucleocytoplasmic mRNA transport: its status in current subcellular biology. In: Bittar EE (ed) Membrane Structure and Function, vol 6. Wiley, New York, pp 43–98

Agutter PS, McArdle HJ, McCaldin B (1976) Evidence of involvement of nuclear envelope NTPase in nucleocytoplasmic translocation of ribonucleoproteins. Nature 263:165–167

Agutter PS, Harris JR, Stevenson I (1977) RNA stimulation of mammalian liver nuclear envelope NTPase. Biochem J 162:671–679

Agutter PS, Cockrill JR, Lavine JE, McCaldin B, Sim RB (1979a) Properties of mammalian liver nuclear envelope NTPase. Biochem J 181:647–658

Agutter PS, McCaldin B, McArdle HJ (1979b) Importance of nuclear envelope NTPase in nucleocytoplasmic RNA transport. Biochem J 182:811–819

Alekseev AB, Yazykov AA, Afanasova LA, Stvolinskii SL (1982) Intracellular transport of nuclear polyadenylated RNA in the process of cell regeneration in *Acetabularia*. Ontogenez 13:655–660

Alonso A, Fischer J, Koenig N, Kinzel V (1981) Structural analysis of HnRNP particles approached by in vivo phosphorylation using exogenous protein kinase and γ-^{32}P-labelled ATP. Eur J Cell Biol 26:208–211

Anderson DM, Richter JD, Chamberlin ME, Price DH, Britten RJ, Smith LD, Davidson EH (1982) Sequence organization of the poly(A) RNA synthesized and accumulated in lampbrush chromosome stage *Xenopus laevis* oocytes. J Mol Biol 155:281–309

Appels R, Tallroth E, Appels DM, Ringertz NR (1975) Differential uptake of protein into the chicken nuclei of HeLa x chicken erythrocyte heterokaryons. Exp Cell Res 92:70–78

Arenstorf HP, Conway GC, Wooley JC, Le Stourgeon WM (1984) Nuclear matrix-like filaments form through artefactual rearrangements of HnRNP particles. J Cell Biol 99:233a

Austerberry CF, Paine PL (1982) *In vivo* distribution of proteins within a single cell. Clin Chem 28:1011–1014

Bachmann M, Schroeder HC, Messer M, Mueller WEG (1984a) Base-specific ribonucleases potentially involved in HnRNA processing and poly(A) metabolism. FEBS Lett 171:25–30

Bachmann M, Bernd A, Schroeder HC, Zahn RK, Mueller WEG (1984b) The role of protein phosphokinase and protein phosphatase during the nuclear envelope NTPase reaction. Biochim Biophys Acta 773:308–316

Bag J (1984) Cytoplasmic mRNA-protein complexes of chicken muscle cells and their role in protein synthesis. Eur J Biochem 141:247–254

Bag J, Sarkar S (1976) Studies on nonpolysomal RNP coding for myosin heavy chains from chick embryonic muscle. J Biol Chem 251:7600–7609

Baglia FA, Maul GG (1983) Nuclear ribonucleoprotein release and NTPase activity are inhibited by antibodies directed against one nuclear matrix glycoprotein. Proc Natl Acad Sci USA 80:2285–2298

Ballantine JEM, Woodland HR (1985) Polyadenylation of histone mRNA in *Xenopus* oocytes and embryos. FEBS Lett 180:224–228

Bannerjee AK (1980) 5' terminal cap structure in eukaryotic messenger RNAs. Microbiol Rev 44:175–205

Barrieux A, Ingraham HA, David DM, Rosenfeld MG (1975) Isolation of messenger-like ribonucleoproteins. Biochemistry 14:1815–1821

Beck JS (1962) The behaviour of certain nuclear antigens in mitosis. Exp Cell Res 28:406–418

Beckmann SL, Chikaraishi DM, Deeb SS, Sueoka N (1981) Sequence complexity of nuclear and cytoplasmic RNA from cloned neuro tumor cell lines and brain sections of the rat. Biochemistry 20:2684–2692

Benyumov AO, Kostomarova AA, Gazaryan KG (1983) Synthesis and transport of nuclear RNA in the early loach *Misgurnus fossilis* embryos obtained by transplantation of nuclei into the egg cells. Zh Obshch Biol 44:694–700

Berezney R (1979) Dynamics of the nuclear protein matrix. In: Busch H (ed) The Cell Nucleus vol 7. Academic Press, London, pp 413–455

Berezney R (1980) Fractionation of the nuclear matrix. I. Partial separation into matrix protein fibrils and a residual ribonucleoprotein fraction. J Cell Biol 85:641–650

Berezney R, Coffey DS (1974) Identification of a nuclear protein matrix. Biochem Biophys Res Commun 60:1410–1417

Bernd A, Schroeder HC, Zahn RK, Mueller WEG (1982a) Modulation of the nuclear envelope NTPase by poly(A) rich mRNA and by microtubule protein. Eur J Biochem 129:43–49

Bernd A, Schroeder HC, Zahn RK, Mueller WEG (1982b) Age-dependence of polyadenylate stimulation of nuclear envelope NTPase. Mech Ageing Dev 20:331–341

Bernd A, Schroeder HC, Leyhausen G, Zahn RK, Mueller WEG (1983) Alteration of activity of nuclear envelope NTPase in quail oviduct and liver in dependence on physiological factors. Gerontology 29:394–398

Bernd A, Altmeyer P, Mueller WEG, Schroeder HC, Holzmann H (1984) Effect of dexamethasone on nuclear envelope nucleoside triphosphatase in fibroblasts 3T3 and melanoma cells MML1. J Invest Dermatol 83:20–22

Berridge MJ (1984) Inositol trisphosphate and diacylglycerol as second messengers. Biochem J 220:3455–360

Berridge MJ, Irvine RF (1984) Inositol trisphosphate, a novel second messenger in cellular signal transduction. Nature 312:315–321

Berrios M, Filson AJ, Blobel G, Fisher PA (1983) A 174-kilodalton ATPase/dATPase polypeptide and a glycoprotein of apparently identical molecular weight are common but distinct components of higher eukaryotic nuclear structural protein subfractions. J Biol Chem 258:13384–13390

Berrios M, Osheroff N, Fischer PA (1985) In situ localization of DNA topoisomerase II, a major polypeptide component of the Drosophila nuclear matrix fraction. Proc Natl Acad Sci USA 82:4142–4146

Beyer AL, Christiensen ME, Walker W, Le Stougeon WM (1977) Identification and characterization of the packaging proteins of core 40S HnRNA particles. Cell 11:127–138

Beyer AL, Miller OL, McKnight SL (1980) Ribonucleoprotein structure in nascent HnRNA is nonrandom and sequence-dependent. Cell 20:75–84

Beyer AL, Bouton AH, Miller OL (1981) Correlation of HnRNP structure and nascent transcript cleavage. Cell 26:155–165

Bhorjee JS, Barclay SL, Wedrychowski A, Smith AM (1983) Monoclonal antibodies specific for tight binding human chromatin antigens reveal structural rearrangements within the nucleus during the cell cycle. J Cell Biol 97:389–396

Bina M, Feldmann RJ, Deeley RG (1980) Could poly(A) align splicing sites? Proc Natl Acad Sci USA 77:1278–1282

Blanchard JM, Ducamp C, Jeanteur P (1975) Endogenous protein kinase activity in nuclear RNP particles from Hela cells. Nature 253:467–468

Blanchard JM, Brunel C, Jeanteur P (1977) Characterization of an endogenous protein kinase activity in RNP structures containing HnRNA from Hela cell nuclei. Eur J Biochem 79:117–131

Blanchard JM, Brunel C, Jeanteur P (1978) Phosphorylation in vivo of proteins associated with HnRNA in Hela cell nuclei. Eur J Biochem 86:301–310

Blobel G (1973) A protein of molecular weight 73,000 bound to the polyadenylate region of eukaryotic messenger RNAs. Proc Natl Acad Sci USA 70:924–928

Bonneau A-M, Darveau A, Sonenberg N (1985) Effect of viral infection on host protein synthesis and mRNA association with the cytoplasmic CSK structure. J Cell Biol 100:1209–1218

Bonner WM (1975a) Protein migration into nuclei. I. Frog oocyte nuclei in vitro accumulate microinjected histones, allow entry to small proteins, and exclude large proteins. J Cell Biol 64:421–430

Bonner WM (1975b) Protein migration into nuclei. II. Frog oocyte nuclei accumulate a class of microinjected oocyte nuclear proteins and exclude a class of microinjected oocyte cytoplasmic proteins. J Cell Biol 64:431–437

Bouvier D, Hubert J, Sève A-P, Bouteille M (1985) Structural aspects of intranuclear matrix disintegration on RNAase digestion of Hela cell nuclei. Eur J Cell Biol 36:323–333

Brasch K (1982) Fine structure and localization of the nuclear matrix in situ. Exp Cell Res 140:161–172

Bryan RN, Hayashi M (1973) Two proteins are bound to most species of polysomal mRNA. Nature New Biol 244:271–274

Burke B, Tooze J, Warren G (1983) A monoclonal antibody which recognizes each of the nuclear lamin polypeptides in mammalian cells. EMBO J 2:361–368

Burkholder G (1983) Silver staining of histone-depleted metaphase chromosomes. Exp Cell Res 147:287–296

Capco DG, Penman S (1983) Mitotic architecture of the cell: the filament networks of the nucleus and cytoplasm. J Cell Biol 96:896–906

Capco DG, Wan KM, Penman S (1982) The nuclear matrix: three-dimensional architecture and protein composition. Cell 29:847–858

Cech TR, Zaug AJ, Grabowski PJ (1981) in vitro splicing of the ribosomal RNA precursor of Tetrahymena: involvement of a guanosine nucleotide in the excision of the intervening sequence. Cell 27:487–496

Cervera M, Dreyfuss G, Penman S (1981) Messenger RNA is translated when associated with the cytoskeletal framework in normal and vesicular stomatitis virus infected Hela cells. Cell 23:113–120

Chakraborty D, Mukherjee AK, Sarkar S, Lee KAW, Darveau A, Sonenberg N (1982) Association of cap binding protein related polypeptides with cytoplasmic ribonucleoprotein particles of chick embryonic muscle. FEBS Lett 149:29–35

Chaly N, Bladon T, Aitchison WA, Setterfield G, Kaplan JG, Brown DL (1983) Redistribution of nuclear matrix antigens during spindle formation. J Cell Biol 97:189a

Chaly N, Bladon T, Setterfield G, Little JE, Kaplan JG, Brown DL (1984) Changes in distribution of nuclear matrix antigens during the mitotic cell cycle. J Cell Biol 99:661–671

Chatterjee NK, Weissbach H (1973) Release of RNA from Hela cell nuclei. Arch Biochem Biophys 157:160–166

Clark TG, Rosenbaum JL (1979) An actin filament matrix in hand-isolated nuclei of Xenopus laevis oocytes. Cell 18:1101–1108

Clawson GA, Smuckler EA (1978) Activation energy for RNA transport from isolated rat liver nuclei. Proc Natl Acad Sci USA 75:5400–5404

Clawson GA, Smuckler EA (1982a) A model for nucleo-cytoplasmic transport of ribonucleoprotein particles. J Theor Biol 95:607–614

Clawson GA, Smuckler EA (1982b) Increased amounts of double-stranded RNA in the cytoplasm of rat liver following treatment with carcinogens. Cancer Res 42:3228–3231

Clawson GA, Koplitz M, Castler-Schechter B, Smuckler EA (1978) Energy utilization and RNA transport: their interdependence. Biochemistry 17:3747–3752

Clawson GA, James J, Woo CH, Friend DS, Moody D, Smuckler EA (1980a) Pertinence of nuclear envelope NTPase activity to transport of RNA from rat liver nuclei. Biochemistry 19:2756–2762

Clawson GA, Koplitz M, Moody DE, Smuckler EA (1980b) Effects of thioacetamide treatment on nuclear envelope NTPase activity and transport of RNA from rat liver nuclei. Cancer Res 40:75–79

Clawon GA, Woo CH, Smuckler EA (1980c) Independent responses of nucleoside triphosphatase EC 3.6.1.4 and protein kinase activities in nuclear envelope following thioacetamide treatment. Biochem Biophys Res Commun 95:1200–1204

Clawson GA, Woo CH, Button J, Smuckler EA (1984) Photo affinity labelling of the major nucleoside triphosphatase of rat liver nuclear envelope. Biochemistry 23:3501–3507

Comerford SA, McLuckie IF, Gorman M, Scott KA, Agutter PS (1985) The isolation of nuclear envelopes. Effects of thiol-group oxidation and of calcium ions. Biochem J 226:95–103

Cook PR, Brazell IA (1975) Supercoils in human DNA. J Cell Sci 19:261–279

Dabauville MC, Franke WW (1982) Karyophilic proteins: polypeptides synthesized in vitro accumulate in the nucleus upon microinjection into the cytoplasm of amphibian oocytes. Proc Natl Acad Sci USA 79:5302–5306

Darnell JE (1982) Variety in the level of gene control in eukaryotic cells. Nature 297:365–371

Davey J, Dimmock NJ, Colman A (1985) Identification of the sequence responsible for the nuclear accumulation of influenza virus nucleoprotein in Xenopus oocytes. Cell 40:667–675

Dazy AC, Borghi H, Puiseux-Dao S (1983) RNA migration in *Acetabularia mediterranea:* effects of cytochalasin B cycloheximide and prolonged dark periods. Plant Sci Lett 30:285–296

De Robertis EM, Longthorne RF, Gurdon JB (1978) Intracellular migration of nuclear proteins in *Xenopus* oocytes. Nature 272:254–256

De Robertis EM, Lienhard S, Parisot RF (1982) Intracellular transport of microinjected 5S and small nuclear RNA. Nature 295:572–577

De Robertis EM (1983) Nucleo-cytoplasmic segregation of proteins and RNAs. Cell 32:1021–1025

Di Maria PR, Kaltwasser G, Goldenberg CJ (1985) Partial purification and properties of a pre-mRNA splicing activity. J Biol Chem 260:1096–1102

Dingwall C, Sharnick SV, Laskey RA (1982) A polypeptide domain that specifies migration of nucleoplasmin into the nucleus. Cell 30:449–458

Dreyer C, Wang YH, Wedlich D, Wyllie CC (1983) Oocyte nuclear proteins in the development of *Xenopus.* In: Hausen P, McLaren A (eds) British Society for Developmental Biology Symposium, Cambridge University Press, pp 285–331

Dreyfuss G, Choi YD, Adam SA (1984) Isolation of the HnRNP complex from vertebrate nuclei. J Cell Biol 99:223a

Drummond DR, McCrae MA, Colman A (1985) Stability and movement of mRNAs and their encoded proteins in *Xenopus* oocytes. J Cell Biol 100:1148–1156

Dubochet J, Morel C, Lebleu B, Merzberg M (1973) Structure of globin mRNA and mRNA-protein particles. Eur J Biochem 36:465–472

Early P, Rogers J, Davis M, Calame K, Bond M, Wall R, Hood L (1980) Two messenger RNAs can be produced from a single immunoglobulin gene by alternative RNA processing pathways. Cell 20:313–319

Earnshaw WC, Honda BM, Laskey RA, Thomas JO (1980) Assembly of nucleosomes: the reaction involving *Xenopus laevis* nucleoplasmin. Cell 21:373–383

Economidis IV, Pederson T (1983a) Structure of nuclear ribonucleoprotein: HnRNA is complexed with a major sextet of proteins in vivo. Proc Natl Acad Sci USA 80:1599–1602

Economidis IV, Pederson T (1983b) In vitro assembly of a pre-messenger ribonucleoprotein. Proc Natl Acad Sci USA 80:4296–4300

Edmonds M, Winters MA (1976) Polyadenylate polymerases. Prog Nucleic Acid Res Mol Biol 17:149–179

Engelhardt P, Plagens U, Zbarsky IB, Filatova LS (1982) Granules 25nm in diameter: basic constituent of the nuclear matrix, chromosome scaffold and nuclear envelope. Proc Natl Acad Sci USA 79:6937–6940

Engelke DR, Shastry BS, Roeder RG (1980) Specific interaction of a purified transcription factor with an internal control region of 5S RNA genes. Cell 19:717–728

Epstein P, Lidsky M, Reddy R, Tan E, Busch H (1982) Identification of three different anti-4S RNA sera associated with autoimmune disease. Biochem Biophys Res Commun 109:548–555

Esumi H, Takahashi Y, Sato S, Nagase S, Sugimura T (1983) A seven-base-pair deletion in an intron of the albumin gene of analbuminemic rats. Proc Natl Acad Sci USA 80:95–99

Feldherr CM (1972) Structure and function of the nuclear envelope. In: Dupraw EJ (ed) Advances in Cell and Molecular Biology, Academic Press, London, pp 273–307

Feldherr CM (1975) The uptake of endogenous proteins by oocyte nuclei. Exp Cell Res 93:411–419

Feldherr CM (1980) Ribosomal RNA synthesis and transport following disruption of the nuclear envelope. Cell Tissue Res 205:157–162

Feldherr CM, Ogburn JA (1980) Mechanisms for the selection of nuclear polypeptides in *Xenopus* oocytes. II: Two-dimensional gel analysis. J Cell Biol 87:589–593

Feldherr CM, Pomerantz J (1978) Mechanism for the selection of nuclear polypeptides in *Xenopus* oocytes. J Cell Biol 78:168–175

Fey EG, Wan KM, Penman S (1984) Epithelial cytoskeletal framework and nuclear matrix/intermediate filament scaffold: three-dimensional organization and protein composition. J Cell Biol 98:1973–1984

Filipowicz W (1978) Functions of the 5' terminal m^{7}G cap in eukaryotic mRNA. FEBS Lett 96:1–11

Filson AJ, Lewis A, Blobel G, Fisher PA (1982) Monoclonal antibodies prepared against the major *Drosophila* nuclear matrix pore complex lamina glycoprotein bind specifically to the nuclear envelope in situ. J Biol Chem 260:3164–3172

Finkel D, Groner Y (1983) Methylations of adenosine residues in pre-messenger RNA are important for formation of late SV-40 mRNA. Virology 131:409–425

Fisher PA, Berrios M, Blobel G (1982) Isolation and characterization of a proteinaceous subnuclear fraction composed of nuclear matrix, peripheral lamina and nuclear pore complexes from embryos of *Drosophila melanogaster*. J Cell Biol 92:674–686

Forbes DJ, Kirschner MW, Newport JW (1983) Spontaneous formation of nucleus-like structures around bacteriophage DNA microinjected into *Xenopus* eggs. Cell 34:13–23

Ford JP, Hsu M-T (1978) Transcription pattern of in vivo labelled late SV40 RNA. J Virol 28:795–801

Franke WW (1970) On the universality of the nuclear pore complex structure. Z Zellforsch Mikrosk Anat 105:405–420

Franke WW (1974) Structure, biochemistry and function of the nuclear envelope. Int Rev Cytol Suppl 4:71–236

Franke WW, Scheer U (1974) Structure and functions of the nuclear envelope. In: Busch H (ed) The Cell Nucleus, vol 1. Academic Press, London, pp 279–347

Franklin RM, Emmons LR, Emmons RP et al. (1984) A monoclonal antibody recognizes an epitope common to an avian specific nuclear antigen and to cytokeratins. J Cell Biochem 24:1–14

Fraser NW, Nevins JR, Ziff E, Darnell JE (1979) The major late adenovirus type-2 transcription unit: termination is downstream from the last poly(A) site. J Mol Biol 129:643–656

French BT, Hanausek-Walazek M, Walszek Z, Schumm DE, Webb TE (1984) Nucleocytoplasmic release of repetitive DNA transcripts in carcinogenesis correlates with a 60 kilodalton cytoplasmic protein. Cancer Lett 23:45–52

Fritzler MJ, Ali R, Tan EM (1984) Antibodies from patients with mixed connective tissue disease react with HnRNP or RNA of the nuclear matrix. J Immunol 132:1216–1222

Fulton AB (1982) How crowded is the cytoplasm? Cell 30:345–347

Furuichi Y, la Fiandra A, Shatkin AJ (1977) 5'-terminal structure and mRNA stability. Nature 266:235–239

Gaedigk R, Oehler S, Koehler K, Setyono B (1985) In vitro reconstitution of messenger RNP particles from globin mRNA and cytosol proteins. FEBS Lett 179:201–207

Gajdardjieva KC, Dabeva MD, Hadjiolov AA (1980) Maturation and nucleocytoplasmic transport of rat liver ribosomal RNA upon D galactosamine inhibition of transcription. Eur J Biochem 104:451–458

Gall JG (1967) Octagonal nuclear pores. J Cell Biol 32:391–402

Gallinaro H, Puvion E, Kister L, Jacob M (1983) Nuclear matrix and HnRNP share a common structural constituent associated with premessenger RNA. EMBO J 2:953–96

Gander ES, Stewart AG, Morel CM, Scherrer K (1973) Isolation and characterization of ribosome-free cytoplasmic mRNP complexes from avian erythroblasts. Eur J Biochem 38:443–452

Gerace L, Blobel G (1980) The nuclear envelope is reversibly depolymerized during mitosis. Cell 19:277–287

Gerace L, Blobel G (1981) Nuclear lamina and the structural organization of the nuclear envelope. Cold Spring Harbor Symp Quant Biol 46:967–978

Gerace L, Blum A, Blobel G (1978) Immuno-cytochemical localization of the major polypeptides of the nuclear pore complex-lamina fraction. J Cell Biol 79:546–566

Gerace L, Ottaviano Y, Kondor-Koch C (1982) Identification of a major polypeptide of the nuclear pore-complex. J Cell Biol 95:826–837

Ghosh S, Pawletz N, Ghosh I (1978) Cytological identification and characterization of the nuclear matrix. Exp Cell Res 111:363–371

Giese G, Wunderlich F (1982) In vitro ribosomal ribonucleoprotein transport: temperature-induced "graded unlocking" of nuclei. J Biol Chem 258:131–135

Goidl JA, Canaani D, Boublik M, Weissbach H, Deckermann H (1975) Polyanion-induced release of polyribosomes from HeLa cell nuclei. J Biol Chem 250:9198–9205

Goldfine ID, Clawson GA, Smuckler EA, Purrello F, Vigneri R (1982a) Action of insulin at the nuclear envelope. Mol Cell Biochem 48:3–14

Goldfine ID, Purrello F, Clawson GA, Vigneri R (1982b) Insulin binding sites on the nuclear envelope: potential relationship to messenger RNA metabolism. J Cell Biochem 20:29–40

Gooderham K, Jeppeson PGN (1983) Chinese hamster metaphase chromosomes isolated under physiological conditions. Exp Cell Res 144:1–14

Grabowski PJ, Padgett RA, Sharp PA (1984) Messenger RNA splicing in vitro: an excised intervening sequence and a potential intermediate. Cell 37:415–427

Granger BL, Lazarides E (1982) Structural associations of synemin and vimentin filaments in avian erythrocytes revealed by immuno-electron microscopy. Cell 30:263–275

Gruss P, Lai C-J, Dhar R, Khoury G (1979) Splicing as a requirement for biogenesis of functional 16S mRNA of SV40. Proc Natl Acad Sci USA 76:4317–4321

Guatelli JC, Porter KR, Anderson KL, Boggs DP (1982) Ultrastructure of the cytoplasmic and nuclear matrices of human lymphocytes observed using high voltage electron microscopy of embedment-free sections. Biol Cell 43:69–80

Gurdon JB (1970) Nuclear transplantation and the control of gene activity in animal development. Proc R Soc Lond Biol Sci B 176:303–314

Gurdon JB, Melton DA (1981) Gene transfer in amphibian cells and oocytes. Annu Rev Genet 15:189–218

Habets WJ, de Rooij DJ, Salden MH, Verhagen AP, Van Eekelen CAG, Van Venrooij WJ, van de Putte LB (1983a) Antibodies against distinct nuclear matrix proteins are characteristic for mixed connective tissue disease. Clin Exp Immunol 54:265–276

Habets WJ, den Brok JH, Boerbooms AMT, van de Putte LBA, Van Venrooij WJ (1983b) Characterization of the SS-B LA antigen in adenovirus infected and uninfected HeLa cells. EMBO J 2:1625–1632

Hadlaczky G, Sumner AT, Ross A (1981) Protein depleted chromosomes. II. Experiments concerning the reality of chromosome scaffold. Chromosoma (Berl) 81:557–567

Hamalainen L, Oikarinen J, Kivirikko KI (1985) Synthesis and degradation of type I procollagen mRNAs in cultured human skin fibroblasts and the effect of cortisol. J Biol Chem 260:720–725

Hancock R, Boulikas T (1982) Functional organization in the nucleus. Int Rev Cytol 79:165–214

Harmon FR, Subrahmanyam CS, Busch H (1985) Interactions of U1-RNP with heterogeneous nuclear RNA in rat Novikoff hepatoma nuclei. Mol Cell Biochem 65:45–55

Havre PA, Evans DR (1983) Disassembly and characterization of the nuclear pore-complex-lamina fraction from bovine liver nuclei. Biochemistry 22:2852–2860

Hazan N, McCauley R (1976) Effect of phenolbarbitone on the nucleocytoplasmic transport of RNA in vitro. Biochem J 156:665–670

Hellmann GM, Chu L-Y, Rhoades RE (1982) A polypeptide which reverses cap analogue inhibition of cell-free protein synthesis. J Biol Chem 257:4056–4062

Herlan G, Eckert WA, Kaffenberger W, Wunderlich F (1980) Isolation and characterization of an RNA-containing nuclear matrix from Tetrahymena macronuclei. Biochemistry 18:1782–1787

Herman RC, Williams JG, Penman S (1976) Heterogeneous nuclear RNP fibers adjacent to poly(A) in steady-state HnRNA in HeLa cells. Cell 7:429–437

Herman RC, Weymouth L, Penman S (1978) Heterogenous nuclear RNP fibers in chromatin – depleted nuclei. J Cell Biol 78:663–674

Hernandez N, Keller W (1983) Splicing of *in vitro* synthesized mRNA precursors in HeLa cell extracts. Cell 35:89–99

Hofer E, Darnell JE (1981) The primary transcription unit of the mouse β-major globin gene. Cell 23:585–593

Hofer E, Hofer-Warbinek R, Darnell JE (1982) Globin RNA transcription: a possible termination site and demonstration of transcriptional control correlated with altered chromatin structure. Cell 29:887–893

Hoffner NJ, Di Bernardino MA (1977) The acquisition of egg cytoplasmic non-histone proteins by nuclei during nuclear reprogramming. Exp Cell Res 108:421–427

Horisberger M, Amos H (1970) Nuclear-cytoplasmic exchange of RNA and protein: effect of temperature and serum. Exp Cell Res 62:467–470

Huez G, Bruck C, Cleuter Y (1981) Translational stability of native and deadenylated rabbit globin mRNA injected into HeLa cells. Proc Natl Acad Sci USA 78:908–911

Igo-Kemenes T, Zachau HG (1978) Domains in chromatin structure. Cold Spring Harbor Symp Quant Biol 42:109–118

Imaizi-Scherrer MT, Maundrell K, Civelli O, Scherrer K (1982) Transcriptional and post transcriptional regulation in duck erythroblasts. Dev Biol 93:126–138

Ishikawa K, Kuroda C, Ogata K (1969) Release of RNP particles containing rapidly-labelled RNA from rat liver nuclei: effect of ATP and some properties of the particles. Biochim Biophys Acta 179:316–331

Ishikawa K, Kuroda C, Ueki M, Ogata K (1970a) Messenger RNP complexes released from rat liver nuclei by ATP. I. Characterization of the RNA moiety of mRNP complexes. Biochim Biophys Acta 213:495–504

Ishikawa K, Ueki M, Nakai K, Ogata K (1970b) Incorporation of nuclear rapidly-labelled RNA into polysomes in the reconstructed system of rat liver. Biochim Biophys Acta 213:542–544

Ishikawa K. Ueki M, Nakai K, Ogata K (1972) Transportation of mRNA from nuclei to polysomes in rat liver cells. Biochim Biophys Acta 259:138–154

Ishikawa K, Sato-Odani S, Ogata K (1978) The role of ATP in the transport of rapidly-labelled RNA from isolated nuclei of rat liver in vitro. Biochim Biophys Acta 521:650–661

Iyer B, Mastro G (1982) A cAMP and cGMP insensitive protein kinase in rat liver nuclear envelopes. J Cell Biol 95:127a

Jackson DA, Cook PR (1985) Transcription occurs at a nucleoskeleton. EMBO J 4:919–926

Jackson DA, McCready SJ, Cook PR (1981) RNA is synthesized at the nuclear cage. Nature 292:552–555

Jacobs H, Birnie GE (1982) Isolation and purification of rat hepatoma nuclei active in the transport of messenger RNA in vitro. Eur J Biochem 121:597–608

Jeffery WR (1977) Characterization of polypeptides associated with mRNA and its poly(A) segment in Ehrlich ascites mRNP. J Biol Chem 252:3525–3532

Jeffery WR (1982) Messenger RNA in the cytoskeletal framework: analysis by *in situ* hybridization. J Cell Biol 95:1–7

Jelinek M, Adesnik M, Salditt M et al. (1973) Further evidence on the nuclear origin and transfer to the cytoplasm of polyadenylic acid sequences in mammalian cell RNA. J Mol Biol 75:515–532

Jelinek W, Goldstein L (1973) Isolation and characterization of some of the proteins that shuttle between nucleus and cytoplasm in *A. proteus*. J Cell Physiol 81:181–197

Jeppeson PGN, Bankier AT, Sanders L (1978) Nonhistone proteins and the structure of metaphase chromosomes. Exp Cell Res 115:293–302

Kalderon D, Richardson WD, Markham AF, Smith AE (1984) Sequence requirements for nuclear location of SV40 large-T antigen. Nature 311:33–38

Kaufmann SH, Shaper JH (1984) Binding of dexamethasone to rat liver *in vivo* and *in vitro:* evidence for two distinct binding sites. J Steroid Biochem 20:699–708

Kaufmann SH, Coffey DS, Shaper JH (1981) Considerations in the isolation of rat liver nuclear matrix, nuclear envelope and pore complex lamina. Exp Cell Res 132:105–123

Kaufmann SH, Gibson W, Shaper JH (1983) Characterization of the major polypeptides of the rat liver nuclear envelope. J Biol Chem 258:2710–2719

Kay RR, Fraser D, Johnston IR (1972) A method for the rapid isolation of nuclear membranes from rat liver. Eur J Biochem 30:145–154

Khalili K, Weinmann R (1984) Actin messenger RNAs in HeLa cells: stabilization after adenovirus infection. J Mol Biol 180:1007–1021

Khawaja JA (1983) Effect of ethanol ingestion on the nucleo-cytoplasmic transport of hepatic RNA. Toxicol Lett 115:199–204

Kindas-Muegge I, Sauermann G (1985) Transport of beta-globin mRNA from nuclei of murine Friend erythroleukemia cells. Eur J Biochem 148:49–54

Kirschner RH, Rusli M, Martin TE (1977) Characterization of the nuclear envelope, pore complex, and dense lamina of mouse liver nuclei by high resolution scanning electron microscopy. J Cell Biol 72:118–132

Kleiman L, Birnie GD, Young BD, Paul J (1977) Comparison of the base-sequence complexity of polysomal and nuclear RNAs in growing Friend erythroleukemia cells. Biochemistry 16:1218–1223

Kletzein RF (1975) Transport of RNA from isolated nuclei — the role of ATP. J Cell Biol 67:217a

Kletzein RF (1980) Nucleocytoplasmic transport of RNA: effect of 3'-deoxy ATP on RNA release from isolated nuclei. Biochem J 192:753–760

Klug A (1983) From macromolecules to biological assemblies. Biosci Rep 3:395–430

Konarska MM, Padgett RA, Sharp PA (1984) Recognition of cap structure in splicing in vitro of mRNA precursors. Cell 38:731–736

Konarska MM, Grabowski PJ, Padgett RA, Sharp PA (1985) Characterization of the branch site in the lariat RNAs produced by splicing mRNA precursors. Nature 313:552–557

Kondor-Koch C, Riedel N, Valentin R, Fasold H, Fischer H (1982) Characterization of an ATPase on the inside of rat liver nuclear envelopes by affinity labelling. Eur J Biochem 127:285–290

Korn ED (1982) Actin polymerization and its regulation by proteins from non muscle cells. Physiol Rev 62:672–737

Korn LJ, Gurdon JB, Price J (1982) Oocyte extracts reactivate developmentally inert *Xenopus* 5S genes in somatic nuclei. Nature 300:354–355

Krachmarov C, Iovcheva C, Hancock R, Dessev GM (1986) Association of DNA with the nuclear lamina in Ehrlich ascites tumor cells. J Cell Biochem 31:59–74

Krachmarov C, Tasheva B, Markov D, Hancock R, Dessev GM (1986) Isolation and characterization of nuclear lamina from Ehrlich ascites tumor cells. J Cell Biochem 30:351–359

Kraemer A, Keller W, Appel B, Luehrmann R (1984) The 5' terminus of the RNA moiety of U1 snRNP particles is required for the splicing of mRNA precursors. Cell 38:299–307

Krainer AR, Maniatis T, Ruskin B, Green MR (1984) Normal and mutant globin pre-mRNAs are faithfully and efficiently spliced in vitro. Cell 36:993–1005

Krohne G, Franke WW, Scheer U (1978a) The major polypeptides of the nuclear pore complex. Exp Cell Res 116:85–102

Krohne G, Franke WW, Ely A, D'Arcy A, Jost E (1978b) Localization of a nuclear envelope-associated protein by indirect immunofluorescence microscopy using antibodies against a major polypeptide from rat liver fractions enriched in nuclear envelope-associated material. Cytobiologie 18:22–38

Krohne G, Stick R, Kleinschmidt JA, Moll R, Franke WW, Hausen P (1982) Immunological localization of a major karyoskeletal protein in nucleoli of oocytes and somatic cells of *Xenopus laevis*. J Cell Biol 94:749–754

Krowczynska A, Yenofsky R, Brawerman G (1985) Regulation of mRNA stability in mouse erythroleukemia cells. J Mol Biol 181:231–239

Kumar A, Pederson T (1975) Comparison of proteins bound to HnRNA and mRNA in HeLa cells. J Mol Biol 96:353–365

Laemmli UK, Cheng SM, Adolph KW, Paulson JR, Brown JA, Baumbach WR (1978) Metaphase chromosome structure. Cold Spring Harbor Symp Quant Biol 42:351–360

Laliberté J-F, Dagenais A, Fillion M, Bibor-Hardy V, Simard R, Royal A (1984) Identification of distinct messenger RNAs for nuclear lamin C and a putative precursor of nuclear lamin A. J Cell Biol 98:980–985

Lang I, Peters R (1984) Nuclear envelope permeability: a sensitive indicator of pore complex integrity. Prog Clin Biol Res 164:377–386

Laskey RA, Earnshaw WC (1980) Nucleosome assembly. Nature 286:763–767

Lasky L, Nozick ND, Tobin AJ (1978) Few transcribed RNAs are translated in avian erythroid cells. Dev Biol 67:23–29

Lebel S, Raymond Y (1984) Lamin B from rat liver nuclei exists both as a lamina protein and as an intrinsic membrane protein. J Biol Chem 259:2693–2696

Lebkowski JS, Laemmli UK (1982a) Evidence for two levels of DNA folding in histone depleted HeLa cell interphase nuclei. J Mol Biol 156:309–324

Lebkowski JS, Laemmli UK (1982b) Non-histone proteins and the long range organisation of HeLa interphase DNA. J Mol Biol 156:325–334

Legname C, Goldstein L (1972) Proteins in nucleocytoplasmic interactions. Exp Cell Res 75:111–121

Lehto VL, Virtaanen I, Kurki P (1978) Intermediate filaments anchor the nuclei in nuclear monolayers of cultured human fibroblasts. Nature 272:175–177

Lemaire M, Bayens W, Baugnet-Mahieu L (1981) On the altered nucleocytoplasmic transport in vitro of rapidly-labeled RNA, in the presence of cytosol or serum from tumor-bearing rats. Biomedicine (Paris) 34:47–53

Lerner MR, Steitz JA (1979) Antibodies to snRNAs complexed with proteins are produced by patients with systemic lupus erythematosus. Proc Natl Acad Sci USA 76:5495–5499

Lerner MR, Boyle JA, Mount SM, Wolin SL, Steitz JA (1980) Are snRNPs involved in splicing? Nature 283:220–224

Lichtenstein AV, Zaboykin MM, Mojseev VL, Shapot VS (1982) Nucleus and cytoplasm; supply and demand. What underlies the flow of genetic information? Subcell Biochem 8:185–250

Littauer UZ, Soreq H (1982) The regulatory function of poly(A) and adjacent 3' sequences in translated RNA. Prog Nucleic Acid Res Mol Biol 27:53–83

Long BH, Huang C-Y, Pogo AO (1979) Isolation and characterization of the nuclear matrix in Friend erythroleukemia cells: chromatin and HnRNA interactions with the nuclear matrix. Cell 18:1079–1090

Longsdon CD, Moessner J, Williams JA, Goldfine ID (1985) Glucocorticoids increase amylase mRNA levels, secretory organelles, and secretion in pancreatic acinar AR42J cells. J Cell Biol 100:1200–1208

Luetzeler J, Verney E, Sidransky H (1979) Nucleolar-cytoplasmic transport of RNA in livers of rats fed on a deficient diet. Exp Mol Pathol 31:261–268

Mangiarotti G, Zuker C, Chisholm RL, Lodish HF (1983) Messengers encoded by genes that are subject to different developmental controls in Dictyostelium display different processing/transit times. Mol Cell Biol 3:1511–1517

Marbaix G, Huez G, Burny A et al. (1975) Absence of polyadenylate segment in globin mRNA accelerates its degradation in Xenopus oocytes. Proc Natl Acad Sci USA 72:3065–3067

Mariman ECM, van Eekelen CAG, Reinders RJ, Berns AJM, Van Venrooij WJ (1982) Adenoviral HnRNA is associated with the host nuclear matrix during splicing. J Mol Biol 154:103–120

Marsden MPF, Laemmli UK (1979) Metaphase chromosome structure: evidence for a radial loop model. Cell 17:849–858

Martin TE, Billings PB, Pullman JM, Stevens BJ, Kinniburgh AJ (1978) Structure of nuclear RNP complexes. Cold Spring Harbor Symp. Quant Biol 42:899–909

Martin TE, Pullman JM, McMullen MD (1980) Structure and function of nuclear and cytoplasmic ribonucleoprotein particles. In: Prescott DM, Goldstein (eds) Cell Biology, vol 4. Academic Press, London, pp 137–174

Mattaj IW, de Robertis EW (1985) Nuclear segregation of U2snRNA requires binding of specific snRNP proteins. Cell 40:111–118

Maul GG (1977) The nuclear and the cytoplasmic pore complex: structure dynamics, distribution and evolution. Int Rev Cytol Suppl 6:75–186

Maul GG (1982) Aspects of a hypothetical nucleocytoplasmic transport mechanism. In: Maul GG (ed) The nuclear envelope and the nuclear matrix, Alan R Liss, New York, pp 1–11

Maul GG, Baglia FA (1983) Localization of a major nuclear envelope protein by differential solubilization. Exp Cell Res 145:285–292

Maundrell K, Maxwell ES, Puvion E, Scherrer K (1981) The nuclear matrix of duck erythroblasts is associated with globin mRNA coding sequences but not with the major proteins of 40S nuclear RNP. Exp Cell Res 136:435–445

Mayrand S, Pederson T (1983) Heat shock alters nuclear RNP assembly in *Drosophila* cells. Mol Cell Biol 3:161–171

McDevitt MA, Imperiale MJ, Ali H, Nevins JR (1984) Requirement of a downstream sequence for generation of a poly(A) addition site. Cell 37:933–999

McDonald JR, Agutter PS (1980) The relationship between polyribonucleotide binding and the phosphorylation and dephosphorylation of nuclear envelope protein. FEBS Lett 116:145–148

McKeon FD, Tuffanell DL, Kobayashi S, Kirschner MW (1984) The redistribution of a conserved nuclear envelope protein during the cell cycle suggests a pathway for chromosome condensation. Cell 35:83–96

McNamara DJ, Racevskis J, Schumm DE, Webb TE (1975) Ribonucleic acid synthesis in isolated rat liver nuclei under conditions of RNA processing and transport. Biochem J 147:193–197

Mednieks MI, Hand AR (1983) Nuclear cyclic AMP-dependent protein kinase in rat parotid acinar cells. Exp Cell Res 149:45–56

Miller TE, Huang CY, Pogo AO (1978) Rat liver nuclear skeleton and RNP complexes containing HnRNA. J Cell Biol 76:675–691

Mills KI, Bell LGE (1982) Protein migration from transplanted nuclei in *Amoeba proteus*. I. Relation to the cell cycle and RNA migration as studied by autoradiography. Exp Cell Res 142:207–214

Mirkovitch J, Mirault M-E, Laemmli UK (1984) Organization of the higher-order chromatin loop: specific DNA attachment sites on nuclear scaffold. Cell 39:223–232

Moffett RB, Webb TE (1981) Regulated transport of messenger RNA from isolated liver nuclei by nucleic-acid binding proteins. Biochemistry 20:3253–3262

Moffett RB, Webb TE (1983) Characterization of a messenger RNA transport protein. Biochim Biophys Acta 740:231–242

Moon RT, Danilchik MV, Hille MB (1982) An assessment of the masked message hypothesis. Dev Biol 93:389–403

Moon RT; Nicosia RF, Olsen C, Hille MB, Jeffery WR (1983) The cytoskeletal framework of sea urchin eggs and embryos: developmental changes in the association of mRNA. Dev Biol 95:447–458

Moore CL, Sharp PA (1985) Accurate cleavage and polyadenylation of exogenous RNA substrate. Cell 41:845–855

Morel C, Kayibanda B, Scherrer K (1971) Proteins associated with globin mRNA in avian erythroblasts. FEBS Lett 18:84–88

Mount SM, Petterson I, Hinterberger M, Karmas A, Steitz JA (1983) The U1 snRNA-protein complex selectively binds a 5' splice site *in vitro*. Cell 33:509–518

Mueller WEG (1976) Endoribonuclease IV. A poly(A)-specific ribonuclease from chick oviduct. Eur J Biochem 70:241–248

Mueller WEG, Arendes J, Zahn RK, Schroeder HC (1978) Control of enzymic hydrolysis of polyadenylate-associated proteins. Eur J Biochem 86:283–290

Mueller WEG, Agutter PS, Bernd A, Bachmann M, Schroeder HC (1985) Role of post-transcriptional events in aging: consequences for gene expression in eukaryotic cells. In: Bergener M, Erminci M, Staehlin HB (eds) Thresholds in aging. 1984 Sandoz Lectures in Gerontology, pp 21–58. Academic Press, London

Mullinger AM, Johnson RT (1980) Packing DNA into chromosomes. J Cell Sci 46:61–86

Munroe SH, Pederson T (1981) Messenger RNA sequences in nuclear RNP particles are complexed with protein as shown by nuclease protection. J Mol Biol 147:437–449

Murnane JP, Painter RB (1983) Altered protein synthesis in ataxia telangi ectasia fibroblasts. Biochemistry 22:1217–1222

Murty CN, Verney E, Sidransky H (1980) Effect of tryptophan on nuclear envelope nucleoside triphosphatase in rat liver. Proc Soc Exp Biol Med 163:155–161

Nakayasu H, Ueda K (1984) Small nuclear RNP complex anchors on the actin filaments in bovine lymphocyte nuclear matrix. Cell Struct Funct 9:317–326

Nevins JR (1983) The pathway of eukaryotic mRNA formation. Annu Rev Biochem 52:441–466

Nevins JR, Darnell JE (1978) Steps in the processing of adenovirus-2 mRNA: poly(A)+ nuclear sequences are conserved and poly(A) addition precedes splicing. Cell 15:1477–1493

Nishizuka Y (1984) The role of protein kinase C in cell surface signal transduction and tumour promotion. Nature 308:693–698

Nudel U, Soreq M, Littauer UZ et al. (1976) Globin mRNA species containing poly(A) segments of different lengths. Eur J Biochem 64:115–121

Olmsted JB (1981) Tubulin pools in differentiating neuroblastoma cells. J Cell Biol 89:418–423

Okada TA, Comings DE (1980) A search for protein cores in chromosomes: is the scaffold an artefact? Am J Hum Genet 32:814–832

Otegui C, Patterson RJ (1981) RNA metabolism in isolated nuclei: processing and transport of immunoglobulin light chain sequences. Nucl Acids Res 9:4767–4781

Ottaviano Y, Gerace L (1985) Phosphorylation of the nuclear lamins during interphase and mitosis. J Biol Chem 260:624–632

Padgett RA, Hardy SF, Sharp PA (1983) Splicing of adenovirus RNA in a cell free transcription system. Proc Natl Acad Sci USA 80:5230–5234

Padgett RA, Konarska MM, Grabowski PJ, Hardy SF, Sharp PA (1984) Lariat RNAs as intermediates and products in the splicing of mRNA precursors. Science 225:898–903

Paine PL (1975) Nucleocytoplasmic movement of fluorescent tracers microinjected into living salivary gland cells. J Cell Biol 66:652–657

Paine PL (1982) Mechanisms of nuclear protein concentration. In: Maul GG (ed) The Nuclear Envelope and the Nuclear Matrix, Alan R Liss, New York, pp 75–84

Paine PL (1985) Nucleocytoplasmic protein distributions: roles of the nuclear envelope and the cytomatrix. In: Clawson GA, Smuckler EA (eds) The Nuclear Envelope and RNA Maturation. UCLA Symposium, vol 26. Alan R Liss, New York, pp 215–231

Paine PL, Feldherr CM (1972) Nucleocytoplasmic exchange of macromolecules. Exp Cell Res 74:81–98

Paine PL, Horowitz SB (1980) The movement of material between nucleus and cytoplasm. In: Prescot DM, Goldstein L (eds) Cell Biology, vol 4. Academic Press, London, pp 299–337

Paine PL, Horowitz SB (1982) Charge as a "signal" in the nucleocytoplasmic distribution of macromolecules. J Cell Biol 95:45a

Paine PL, Moore LC, Horowitz SB (1975) Nuclear envelope permeability. Nature 254:109–114

Paine PL, Austerberry CF, Desjarlais LJ, Horowitz SM (1983) Protein loss during nuclear isolation. J Cell Biol 97:1240–1242

Palayoor T, Schumm DE, Webb TE (1981) Transport of functional messenger RNA from liver nuclei in a reconstituted cell-free system. Biochim Biophys Acta 654:201–210

Patzelt E, Blaas D, Kuechler E (1982) Cap binding proteins associated with the nucleus. Nucl Acids Res 11:5821–5835

Paul J (1982) Transcriptional control during development. Biosci Rep 2:63–76

Pederson T (1974) Proteins associated with HnRNA in eukaryotic cells. J Mol Biol 83:163–183

Pederson T (1983) Nuclear RNA-protein interactions and mRNA processing. J Cell Biol 97:1321–1326

Pederson T, Davis NG (1980) Messenger RNA processing and nuclear structure: isolation of nuclear ribonucleoprotein particles containing β-globin mRNA precursors. J Cell Biol 87:47–54

Perry RP, Kelley DE (1968) Messenger RNA-protein complexes and newly synthesized ribosomal subunits: analysis of free particles and components of polyribosomes. J Mol Biol 35:37–59

Peters R (1983a) Fluorescence microphotolysis. Diffusion measurements in single cells. Naturwissenschaften 70:294–302

Peters R (1983b) Nuclear envelope permeability measured by fluorescence microphotolysis of single liver cell nuclei. J Biol Chem 258:11427–11429

Pieck ACM, van der Velden HMW, Rijken AAM, Neis JM, Wanka F (1985) Protein composition of the chromosomal scaffold and interphase nuclear matrix. Chromosoma (Berl) 91:137–144

Pouchelet M, Anteunis A, Gansmuller A, Essien M (1984) Nuclear matrix *in situ*. Biol Cell 52:89a

Proudfoot NJ, Brownlee GG (1976) 3'-noncoding sequences in eukaryotic mRNA. Nature 263:211–214

Pullman JM, Martin TE (1983) Reconstitution of nucleoprotein complexes with mammalian HnRNP core proteins. J Cell Biol 97:99–111

Purrello F, Vigneri R, Clawson GA, Goldfine ID (1982) Insulin stimulation of nucleoside triphosphatase activity in isolated nuclear envelopes. Science 216:1005–1007

Purrello F, Burnham DB, Goldfine ID (1983) Insulin regulation of protein phosphorylation in isolated rat liver nuclear envelopes: potential relationship to messenger RNA metabolism. Proc Natl Acad Sci USA 80:1189–1193

Rao MVN, Prescott DM (1967) Return of RNA into the nucleus after mitosis. J Cell Biol 35:109a

Raskas HJ, Bhaduri S (1973) Poly(A) sequences in adenovirus RNA released from isolated nuclei. Biochemistry 12:920–925

Raskas HJ, Rho J-C (1973) ATP requirement for release of adenovirus mRNA from isolated nuclei. Nature New Biol 245:47–49

Rechsteiner M, Kuehl L (1979) Microinjection of the nonhistone chromosomal protein HMG1 into bovine fibroblasts and HeLa cells. Cell 16:901–908

Richardson JCW, Maddy AH (1980) The polypeptides of rat liver nuclear envelope. I. Examination of nuclear pore complex polypeptides by solid-state lactoperoxidase labelling. J Cell Sci 43:253–267

Riedel N, Fasold H, Bachmann M, Prochnow D (1987) Permeability measurements in closed vesicles from rat liver nuclear envelopes. Proc Natl Acad Sci USA 84:3540–3544

Rinke J, Steitz JA (1982) Precursor molecules of both human 5S ribosomal RNA and transfer RNAs are bound by a cellular protein reactive with anti-La lupus antibodies. Cell 29:149–159

Roberts K, Northcote DH (1970) Structure of the nuclear pore in higher plants. Nature 228:385–387

Rokowski RJ, Sheehy J, Cutroneo KR (1981) Glucocorticoid-mediated selective reduction of functioning collagen mRNA. Arch Biochem Biophys 210:74–81

Rose KM, Jacob ST (1980) Phosphorylation of nuclear poly(A) polymerase by protein kinase: mechanism of enhanced poly(A) synthesis. Biochemistry 19:1472–1476

Rosenfeld MG, Amara SG, Evans RM (1984) Alternative RNA processing: determining neuronal phenotype. Science 225:1315–1320

Ross DA, Yen RW, Chae CB (1982) Association of globin RNA and its precursors with the chicken erythroblast nuclear matrix. Biochemistry 21:764–771

Ruskin B, Krainer AR, Maniatis T, Green MR (1984) Excision of an intact intron as a novel lariat structure during pre-mRNA splicing in vitro. Cell 38:317–331

Ryffel GU (1976) Comparison of cytoplasmic and nuclear poly(A)-containing RNA sequences in *Xenopus* liver cells. Eur J Biochem 62:417–423

Salden MHL, van Eekelen CAG, Habets WJA, Vierwinden G, van de Putte LBA, van Venrooij WJ (1982) Anti nuclear matrix antibodies in mixed connective tissue disease. Eur J Immunol 12:783–786

Salditt-Georgieff M, Darnell JE (1982) Further evidence that the majority of primary nuclear RNA transcripts in mammalian cells do not contribute to mRNA. Mol Cell Biol 2:701–707

Samarina OP, Lukanidin EM, Molnar J, Georgiev GP (1968) Structural organization of nuclear complexes containing DNA-like RNA. J Mol Biol 33:251–1263

Schneider JH (1959) Factors affecting the release of nuclear RNA from the nucleus *in vitro*. J Biol Chem 234:2728–2732

Schroeder HC, Dose K, Zahn RK, Mueller WEG (1980a) Isolation and characterization of the novel poly(A) and poly(U) degrading endoribonuclease V from calf thymus. J Biol Chem 225:5108–5112

Schroeder HC, Zahn RK, Dose K, Mueller WEG (1980b) Purification and characterization of a poly(A)-specific exoribonuclease from calf thymus. J Biol Chem 225:4535–4538

Schroeder HC, Shenk P, Baydown H, Wagner KA, Mueller WEG (1983) Occurrence of short-sized oligo (A) fragments during course of cell cycle and ageing. Arch Gerontol Geriat 2:349–360

Schroeder HC, Rottmann M, Bachmann M, Müller WEG (1986) Purification and characterization of the major NTPase from rat liver nuclear envelopes. J Biol Chem 261:663–668

Schumm DE, Webb TE (1978) Effect of adenoisine-3', 5'-monophosphate and guanosine-3', 5'-monophosphate on RNA release from isolated nuclei. J Biol Chem 253:8513–8517

Schumm DE, Webb TE (1981) Insulin modulated transport of RNA from isolated liver nuclei. Arch Biochem Biophys 210:275–279

Schumm DE, Webb TE (1982) Site of action of colchicine on RNA release from liver nuclei. Biochem Biophys Res Commun 105:375–382

Schumm DE, Morris HP, Webb TE (1973) Cytosol-modulated transport of messenger RNA from isolated nuclei. Cancer Res 33:1821–1828

Schumm DE, Hanausek-Walasek M, Yannarell A, Webb TE (1977) Changes in nuclear RNA transport incident to carcinogenesis. Eur J Cancer 13:139–147

Schwartz H, Darnell JE (1976) The association of protein with poly(A) of HeLa cell mRNA: evidence for a "transport" role of a 75,000 molecular weight polypeptide. J Mol Biol 104:833–851

Schwartzbauer JE, Tamkun JW, Lemischka IR, Hynes RO (1983) Three different fibronectin messenger RNAs arise by alternative splicing within the coding region. Cell 35:421–431

Schweiger A; Kostka G (1984) Concentration of particular high molecular mass phosphoproteins in rat liver nuclei and nuclear matrix decreases following inhibition of RNA synthesis with alpha-amanitin. Biochim Biophys Acta 782:262–268

Shatkin AJ (1976) Capping of eukaryotic mRNAs. Cell 9:645–653

Shatkin AJ (1985) mRNA cap binding proteins: essential factors in initiating translation. Cell 40:223–224

Shearer RW (1974) Specificity of chemical modification of RNA transport by liver carcinogens in the rat. Biochemistry 13:1764–1769

Shearer RW (1977) Differences and similarities in RNA regulation between tumors induced by chemicals and by DNA and RNA viruses. Vestn Akad Med Nauk SSSR 3:64–72

Shelton KR (1985) The nuclear lamins. In: Berezney R (ed) The Nuclear Matrix. Plenum, New York (in press)

Shelton KR, Higgins LL, Cochran DL, Ruffulo JJ, Egle PM (1980) Nuclear lamins of erythrocyte and liver. J Biol Chem 255:10978–10983

Shimotahno K, Kodama Y, Hashimoto J, Miura KI (1977) Importance of 5' terminal blocking structure to stabilize mRNA in eukaryotic protein synthesis. Proc Natl Acad Sci USA 74:2734–2738

Skaer RJ, Whytock S (1977) The fixation of nuclei in glutaraldehyde. J Cell Sci 27:13–22

Skoglund U, Andersson K, Bjorkroth B, Lamb MM, Daneholt B (1983) Visualisation of the formation and transport of a specific HnRNP particle. Cell 34:847–855

Smart-Nixon S, Schumm DE, Webb TE (1983) Organ and species specificity of the messenger RNA transport factor. Comp Biochem Physiol 75:655–670

Smith CD, Wells WW (1983) Phosphorylation of rat liver nuclear envelopes. I: Characterization of in vitro protein phosphorylation. J Biol Chem 258:9360–9367

Smith CD, Wells WW (1984) Solubilization and reconstitution of a nuclear envelope associated ATPase; synergistic activation by RNA and polyphosphoinositides. J Biol Chem 259:11890–11894

Smith DE, Fisher PA (1984) Identification, developmental regulation and response to heat shock of two antigenically related forms of a major nuclear envelope protein in *Drosophila* embryos. J Cell Biol 99:20–28

Smith DWE, McNamara AL (1971) Specialization of rabbit reticulocyte tRNA content for hemoglobin synthesis. Science 171:577–579

Smuckler EA, Koplitz M (1974) Thioacetamide-induced alterations in nuclear RNA transport. Cancer Res 34:827–838

Smuckler EA, Koplitz M (1976) Poly(A) content and electrophoretic behavious of in vitro released RNAs in chemical carcinogenesis. Cancer Res 36:881–888

Sommerville J (1981) Immunolocalization and structural organization of nascent RNP. In: Busch H (ed) The Cell Nucleus vol 8. Academic Press, London, pp 1–57

Sonenberg N, Shatkin AJ (1977) Reovirus mRNA can be covalently crosslinked via the 5' cap to proteins in initiation complexes. Proc Natl Acad Sci USA 74:4288–4292

Spirin AS (1969) Informosomes. Eur J Biochem 10:20–35

Stacey DW, Allfrey VG (1984) Microinjection studies of protein transit across the nuclear envelope of human cells. Exp Cell Res 154:283–292

Staufenbiel M, Deppert W (1982) Intermediate filament systems are collapsed on to the nuclear surface after isolation of nuclei from tissue culture cells. Exp Cell Res 138:207–214

Steitz JA, Kamen R (1981) Arrangement of 30S HnRNP on polyoma virus late nuclear transcripts. Mol Cell Biol 1:21–34

Stevenin J, Gallinaro-Matringe H, Gattoni R, Jacob M (1977) Complexity of the structure of particles containing HnRNA as demonstrated by RNAase treatment. Eur J Biochem 74:589–602

Stevens BJ, Swift H (1966) RNA transport from nucleus to cytoplasm in *Chironomus* salivary glands. J Cell Biol 31:55–57

Strub K, Galli G, Busslinger M, Birnsteil ML (1984) The cDNA sequences of the sea urchin U7 snRNA suggest specific contacts between the histone mRNA precursor and U7 RNA during RNA processing. EMBO J 3:2801–2807

Stuart SE, Rottman FM, Patterson RJ (1975) Nuclear restriction of nucleic acids in the presence of ATP. Biochem Biophys Res Commun 62:439–447

Stuart SE, Clawson GA; Rottman FM, Patterson RJ (1977) RNA transport in isolated myeloma nuclei: transport from membrane-denuded nuclei. J Cell Biol 72:57–66

Stubblefield E (1973) The structure of mammalian chromosomes. Int Rev Cytol 35:1–60

Sugawa H, Uchida T (1985) Inhibition of RNA nucleocytoplasmic translocation by antinucleus antibody. Biochem Biophys Res Commun 127:864–870

Thomas PS, Shepherd JH, Mulvihill ER, Palmiter RD (1981) Isolation of a nuclear ribonucleoprotein fraction from chick oviduct containing ovalbumin mRNA sequences. J Mol Biol 150:143–166

Tobian JA, Castano JG, Zasloff MA (1984) RNA nuclear transport in *Xenopus laevis* oocytes: studies with human initiator tRNA-met point mutants. J Cell Biol 99:232a

Traschel H, Sonenberg N, Shatkin AJ et al. (1980) Purification of a factor that restores translation of vesicular stomatitis virus mRNA in extracts from poliovirus-infected HeLa cells. Proc Natl Acad Sci USA 77:770–774

Treisman R, Orkin SN, Maniatis T (1983) Specific transcription and RNA splicing defects in five cloned β-thalassemic genes. Nature 302:591–596

Tsanev RG, Djondjurov LP (1982) Ultrastructure of free RNP complexes in spread mammalian nuclei. J Cell Biol 94:662–666

Tsiapalis CM, Dorson JW, Bollum FJ (1975) Purification of terminal riboadenylate transferase from calf thymus gland. J Biol Chem 250:4486–4496

Unwin PN, Milligan RA (1982) A large particle associated with the perimeter of the nuclear pore complex. J Cell Biol 93:63–75

Van Eekelen CAG, Rieman T, van Venrooij WJ (1981) Specificity of the interaction of HnRNA and mRNA with proteins as revealed by *in vivo* cross-linking. FEBS Lett 130:223–226

Van Eekelen CAG, van Venrooij WJ (1981) HnRNA and its attachment to a nuclear matrix. J Cell Biol 88:554–563

Van Venrooij WJ, Sillekens PTG, van Eekelen CAG, Reinders RJ (1981) On the association of mRNA with the cytoskeleton in uninfected and adenovirus infected human KB cells. Exp Cell Res 135:79–92

Van Voorthuizen WF, Dinsart C, Flavell RA, Vijlder JJM de, Vassart G (1978) Abnormal cellular localization of thyroglobulin mRNA associated with hereditary congenital goitre and thyroglobulin deficiency. Proc Natl Acad Sci USA 75:74–78

Villarreal LP, Whyte RT (1983) A splice junction deletion deficient in the transport of RNA does not polyadenylate nuclear RNA. Mol Cell Biol 3:1381–1388

Vogelstein B, Pardoll DM, Coffey DS (1980) Supercoiled loops and DNA replication. Cell 22:79–85

Vorbrodt A, Maul GG (1980) Cytochemical studies on the relation of nucleoside triphosphatase activity to ribonucleoproteins in isolated rat liver nuclei. J Histochem Cytochem 28:27–35

Walaszek Z, Hanausek-Walaszek M, Schumm DE, Webb TE (1983) An onco-fetal 60 kilodalton protein in the plasma of tumor-bearing and carcinogen-treated rats. Cancer Lett 20:277–282

Wallace JC, Edmonds M (1983) Polyadenylated nuclear RNA contains branches. Proc Natl Acad Sci USA 80:950–954

Webb NR, Chari RVJ, Kozarich JW, Rhoades RE (1984) Purification of the mRNA cap-binding protein using a new affinity medium. Biochemistry 23:177–181

Webb TE, Schumm DE, Palayoor T (1981) Nucleocytoplasmic transport of mRNA. In: Busch H (ed) The Cell Nucleus, vol 9. Academic Press, London, pp 199–247

Weck PK, Johnson TC (1976) Nuclear-cytosol interactions that modulate RNA synthesis and transcript size in mouse brain nuclei. J Neurochem 27:1367–1374

Weck PK, Johnston TC (1978) Nuclear-cytosol interactions that facilitate release of RNA from mouse brain nuclei. J Neurochem 30:1057–1065

Wickens MP, Gurdon JB (1983) Post-transcriptional processing of SV40 late transcripts in injected frog oocytes. J Mol Biol 163:1–26

Wieben ED, Madore SJ, Pederson T (1983) Protein binding sites are conserved in U1 snRNA from insects and mammals. Proc Natl Acad Sci USA 80:1217–1220

Wieringa B, Meyer F, Reiser J, Weissmann C (1983) Unusual sequence of cryptic splice sites in the β-globin gene following inactivation of an authentic 5' splice site by site-directed mutagenesis. Nature 301:38–43

Wieringa B, Hofer E, Weissmann C (1984) A minimal intron length but no specific internal sequence is required for splicing the large rabbit β-globin intron. Cell 37:915–925

Wojtkowiak Z, Kuhl DM, Briggs RC, Hnilica LS, Stein J, Stein GS (1982) A nuclear matrix antigen in HeLa and other human malignant cells. Cancer Res 42:4546–4552

Wolosewick JJ, Porter KR (1976) Microtrabecular lattice of the cytoplasmic ground substance. J Cell Biol 82:114–139

Wu RS, Warner JR (1971) Cytoplasmic synthesis of nuclear proteins. J Cell Biol 51:643–652

Wunderlich F, Herlan G (1977) A reversibly contractile nuclear matrix: its isolation, structure and composition. J Cell Biol 73:271–278

Wunderlich F, Berezney R, Kleinig H (1976) The nuclear envelope. In: Chapman D, Wallach DFH (eds) Biological Membranes Academic Press, London, pp 241–333

Wunderlich F, Giese G, Herlan G (1984a) Thermal down-regulation of exportable rRNA in nuclei. J Cell Physiol 120:211–218

Wunderlich F, Giese G, Speth V (1984b) Thermal diminution and augmentation of the retention of transportable rRNA in nuclear envelope-free nuclei. Biochim Biophys Acta 782:187–194

Yamaizumi M, Uchida T, Okada Y, Furusawa M, Mitsui H (1978) Rapid transfer of non-histone chromosomal proteins to the nucleus of living cells. Nature 273:782–784

Yannarell A, Awad A-B (1982) The effect of alteration of nuclear lipids on mRNA transport from isolated nuclei. Biochem Biophys Res Commun 108:1056–1060

Young RA, Hagenbuechle O, Schibler U (1981) A single mouse α-amylase gene specifies two different tissue-specific messenger RNAs. Cell 23:451–458

Yu L-C, Racevskis J, Webb TE (1972) Regulated transport of ribosomal subunits from regenerating rat liver nuclei in a cell-free system. Cancer Res 32:2314–2321

Zasloff M (1983) tRNA transport from the nucleus in a eukaryotic cell: carrier-mediated translocation process. Proc Natl Acad Sci USA 80:6436–6440

Zbarskii IB, Peskin AV (1982) Superoxide dismutase activity and the formation of super oxide radicals by membranes in tumor and normal tissues. Vestn Akad Med Nauk SSSR 10:24–28

Zeevi M, Nevins JR, Darnell JE (1982) Newly formed mRNA lacking poly(A) enters the cytoplasm and the polyribosomes but has a shorter half-life in the absence of poly(A). Mol Cell Biol 2:517–525

Zeller R, Nyffenegger T, de Robertis EM (1983) Nucleocytoplasmic distribution of snRNPs and stockpiled snRNA-binding proteins during oogenesis and early development in *Xenopus laevis*. Cell 32:425–434

Zieve GW (1984) SnRNP in the cytoplasm. J Cell Biol 99:232a

Zumbe A, Staehli C, Traschel H (1982) Association of a 50K molecular weight cap binding protein with the cytoskeleton in baby hamster kidney cells. Proc Natl Acad Sci USA 79:2927–2931

Prebiotic Evolution and the Origin of Life: Chemical and Biochemical Aspects

Klaus Dose[1, 2]

A. The Object and Its History

Evolution, as the term is used here, signifies any development or change adapting to the environment. Chemical evolution connotes changes of chemical substances, it thus signifies that changes occur fundamentally in the molecules. Frequently "chemical evolution" is used synonymously for "abiotic" or "prebiotic formation" of organic molecules in a cosmic system, usually on the prebiotic (or primitive) Earth. It is then assumed that the organic molecules were formed from the constituents of the primitive atmosphere, hydrosphere, and — in part — lithospere.

"Molecular evolution" has a broader meaning than chemical evolution. The latter term also covers self-assembly into higher structures (such as membranes, protocells, cell-like systems, protocellular organelles) and the subsequent evolution of protocells to the first living cells ("Urzellen" or progenotes).

From this stage on, the evolution of living cells and organisms to the multiplicity of contemporary systems is called Darwinian evolution. Several authors call the evolution of protocells to the first living cells also proto-Darwinian evolution.

Cosmologists have used the term evolution when referring to the development of stars, solar systems, and milky ways. Nuclear chemists have applied the term evolution when describing the formation or conversion of elements in the interior of stars.

As can be seen, the term evolution has a broad meaning and is used to describe basically different processes.

As early as during the winter semester of 1865/66 the German biologist Ernst Haeckel (1834–1919) proposed in his lectures on Darwinism an evolutionary sequence in order to explain the origin of the first cells by self-organization and selection (Haeckel 1868). Haeckel's pioneering ideas constituted a departure from the then prevailing view that the origin of life could not be a subject of scientific research, because it was supposed to have existed since eternity in an eternal universe. With his evolutionary concept he stirred up vehement emotions among scientists and clergymen during the second half of the 19th century. Oddly, his ideas had become forgotten by 1924 when A.I. Oparin (1924) published bis evolutionary view of the origin of life.

[1]Institut für Biochemie, Johannes Gutenberg-Universität, Becherweg 30, D-6500 Mainz, FRG
[2]To Prof. Dr. Theodor Wieland on the occasion of this 75th birthday

The view that contemporary life arose from primitive life by evolution is not a modern idea. It is found in a rudimentary form already within the framework of various antique philosophies. According to Empodocles (495?–435? B.C.), living organisms arise by spontaneous generation; but many of the organisms primarily generated are malformed beings that perish early whereas only the fitter organisms are selected for further development. Also the philosophy of Aristotle (384–322 B.C.) is governed by developmental ideas. A crucial thesis is that coming into being means development. The potential of coming into being is the force that materializes, forms and completes all matter, alive or dead.

Between 1860 and 1870 Harvey's thesis (1651) "Omne vivum ex ovo" (all life originates from an egg) had become widely accepted. Harvey, however, had only concluded that higher life (e.g. vertebrates) originates ultimately from an egg. He did not state that this generalization principally excluded spontaneous generation of lower forms of life, e.g. insects. Stimulated by Harvey's thesis, Francesco Redi (1626–1697) (Redi 1909, posthumous), for example, disproved the spontaneous generation of "worms" (maggots) in decomposing meat. Redi clearly demonstrated that the maggots originated from the eggs that had been previously deposited on the meat by flies. These experiments were conducted in a very thoughtful manner. Like Pasteur's experiments of 1859–1862 (see below) they deserve to be regarded as monuments in the history of the biosciences. Details of these and related experiments and the history of the object have been repeatedly described (Oparin 1957; Kenyon and Steinman 1969; Fox and Dose 1977; see also the historical review by Farley 1977).

By the middle of the 19th century Harvey's thesis had been confirmed by numerous experiments. The last outpost of the doctrine of spontaneous generation was finally the origin of bacteria and other microorganisms. By his simple and elegant experiments Pasteur (1922, posthumous) disproved the spontaneous generation of microorganisms and demonstrated that microorganisms only develop in a sterile bouillon after it has become contaminated by living germs from the exterior. In his triumphal lecture at the Sorbonne in 1864 he therefore stated (translated from Pasteur 1922, posthumous):

"Never will the doctrine of spontaneous generation recover from the mortal blow struck by this simple experiment". In the same speech at the Sorbonne, he said, "There is the question of so-called spontaneous generation. Can matter organize itself? In other words, are there beings that can come into the world without parents, without ancestors? That is the question to be resolved."

In equating spontaneous generation to a self-organizing act, Pasteur directed attention to the basic issue of life derived from life versus the origin of life from an appropriate non-living precursor or precursors. But Pasteur, at that time, had not evaluated an evolutionary origin of primitive life from even more primitive precursors when he stated, "No, today there is no circumstance known under which one could affirm that microscopical beings have come into the world without germs, without parents resembling themselves."

Darwin in 1859 avoided discussing the question of spontaneous generation. He suggested that all forms of life that ever existed on Earth evolved from a common ancestor. Then he expressed that this common ancestor of all life was created by a divine act. Again in 1863 Darwin rejected the idea that the question of the origin of life could be rationally analyzed. In a letter to Hooker (edited by F. Darwin 1896) he wrote, "It is mere rubbish thinking at the present of the origin of life — one might as will think of the origin of matter". But later, perhaps impressed by Haeckel's thesis on the origin of life, Darwin in 1871 (Darwin 1959, posthumous) no longer rejected thinking of an evolutionary origin of life when he expressed, "It is often said that all the conditions for the first production of a living organism are now present which could ever have been present. But if (and oh! what a big if!) we could conceive in some warm little pond, with all sorts of ammonia and phosphoric salts, light, heat, electricity etc. present, that a protein compound was chemically formed ready to undergo still more complex changes, at the present day such matter would be instantly devoured or absorbed, which could not have been the case before living creatures were formed".

By 1865 many scientists believed that life is eternal. Thus the origin of life could not be a subject of serious research. Richter (1865) stated that germs of life are present everywhere in the eternal and infinite cosmos (Panspermia). Whenever these germs reach an appropriate celestial body, e. g. a planet like the Earth, a wealth of living organisms would emerge. Varying Harvey's thesis, Richter stated, "Omne vivum ab aeternitate e cellula" (all life since eternity originates from a cell). Today, however, we realize, that Richter's germs of life (e. g. bacterial spores) would have to travel at least several hundred thousand years to reach our Earth from a planet outside our solar system. Even if well packed inside meteoritic or cometary materials and thus protected from UV light as well as from most ionizing radiations, from the effects of space vacuum, and collisions with micrometeorites, during its long journey such a spore would very likely be hit by an energy-rich particle (HZE particle) of cosmic radiation and inactivated (see e. g. Crick and Orgel 1973).

B. Modern Hypotheses, General Objections and Supports

The modern era of the field began when Oparin in 1924 for the first time published his concept on the origin of life. But Oparin actually revived Haeckel's 19th century concept (Oparin 1924; Haeckel 1868). Certainly, Oparin could be more specific than Haeckel, whose detailed views today sound quite naive. The gap between the idealized concept and the known experimental facts, however, was also tremendous in 1924 and still continues to be so. It took more than 30 years until some of Oparin's (and Haeckel's) ideas on the spontaneous formation of biologically significant molecules from "inorganic" precursors could be verified in the laboratory. The history of this era of experimentation has been detailed elsewhere (see e. g. Oparin 1957; Fox and Dose 1977; Kenyon and Steinman 1969; Farley 1977).

In Table 1 the current hypotheses on the origin of life are summarized.

Table 1. Hypotheses on the origin of life

Divine creation:	most religions
Panspermia:	
Spontaneous:	Anaximander (500 B.C.); H. Richter (1865), S. Arrhenius (1908)
Directed:	F.H.C. Crick and L.E. Orgel (1973)
Evolution and self-assembly:	
General:	Aristotle (384–322 B.C.) E. Haeckel (1868); A.I. Oparin (1924)
DNA-first:	H.J. Muller (1966); F.H. Crick (1968)
Proteins-first:	E. Haeckel (1868); S.W. Fox (1968)
Coecolution:	M. Calvin (1969)

The principal objections to these hypothese are briefly summarized in Table 2.

Prior to a more specific and detailed discussion of the current concepts in prebiotic evolution a brief summary of the major facts that are in support of an origin of life by evolution and self-assembly is presented in Table 3.

Table 2. Principal objections to the hypotheses on the origin of life

Divine creation:	Contradicts natural laws.
Panspermia (from outside the solar system):	
General:	Provides no explanation for the origin of life in the universe
Spontaneous:	Spores, even within meteorites, will be killed by cosmic radiation during journey of several hundred thousands of years (assumed dose rate: 50 to 100 rem per year).
Evolution and self-assembly:	
General:	Provides no satisfactory explanation for the origin of chirality and the evolution of specific structure-function relationships in cells. No characterization of evolutionary precursors of modern cells available.
DNA-first:	Origin of first replicable polynucleotides unknown
Proteins-first:	Flow of information from prebiotic polypeptides (nonrandom) to first informational polynucleotides not satisfactory established.
Coevolution:	All objections combined.

Table 3. Major facts in support of an origin of life by evolution and self-assembly

General:	1. Abundance of bioelements in the universe.
	2. Formation of a significant variety of organic molecules by simulated prebiotic reactions and in cosmic systems (e.g. interstellar clouds).
	3. Thermodynamic feasibility of evolution and self-organization.

C. The Oparin Hypothesis and the Simulation Experiments

In modern times the evolutionary hypothesis on the origin of life was put forth by Oparin (1924, 1957, 1964).

In its fully developed form the hypothesis is based on four prerequisites:

1. The prebiotic atmosphere was reducing and contained methane, ammonia, hydrogen, and water.
2. This atmosphere was exposed to various forms of energy (e. g. electric discharges, solar radiation, and volcanic heat) which led to the production of organic compounds.
3. These compounds accumulated in the primitive "hydrosphere that reached the consistency of a hot dilute soup".
    4. In this "soup" the first forms of life evolved spontaneously.

This concept has received much criticism. It is now generally accepted that the primitive atmosphere was built up by volcanic outgassing. It was reducing to non-oxidizing and probably consisted of carbon dioxide, water vapour, nitrogen, hydrogen, carbon monoxide and hydrogen sulphide. Its composition has most likely changed during the various phases of chemical evolution from reducing (rich in hydrogen compounds) to non-oxidizing (rich in carbon dioxide and nitrogen). But it never contained the large amounts of free and bound hydrogen that were used in the first Miller-Urey experiments (Miller 1953; Miller et al. 1976). Critical reviews of this revised concept have been presented in various monographs on the origin of life. (see e.g. Fox and Dose 1977; Folsome 1979; Thaxton et al. 1984).

The Miller-Urey experiments and their sequels have taught us much about the chemistry of gas mixtures that are activated by the input of various forms of energy. But the main products have been ill-defined polymers that are more related to tar or to the organic compounds in carbonaceous meteorites (Lawless and Peterson 1976) than to life. Only a limited number of bioorganic compounds have been formed in significant amounts. Among these are the amino acids glycine and D-, L-alanin [yields based on carbon (0.46%–2.1%) and (0.08%–1.8%) respectively] and the fatty acids formic acid (0.4%–3.9%), acetic acid (0.5%–0.7%), propionic acid (0.2%–0.6%) as well as glycolic acid (0.2%–1.9%) and D, L-lactic acid (0.03%–1.8%).

Many more organic compounds of biological relevance have been identified although they were formed with extremely low yields (less than 0.1%). These comprise several hundred different amino acids (total yield 1.9%) including the D-, L forms of almost all proteinous amino acids, a variety of fatty acids, complex mixtures of hydrocarbons, heterocyclic N-compounds including all nucleic acid bases and many other small compounds, but all in racemic mixture if they contain chiral C-atoms (Fox and Dose 1977; Folsome 1979; Miller and Orgel 1974). Substantial amounts of sugars, including D,L-ribose, have never been produced in realistic prebiotic simulation experiments. These negative results suggest that nucleosides, nucleotides or even polynucleotides were not present in significant amounts on the prebiotic Earth (see

Shapiro 1984, 1986 a, b). The conclusion is that either nucleotides and nucleic acids were first formed by early biosynthetic processes or — less likely in view of 30 years of laboratory research on chemical evolution — that the prebiotic routes to nucleotides have simply remained undiscovered. One problem related to this difficulty has been mentioned earlier: We do not know how the prebiotic conditions really were on the early Earth.

It has often been argued that because Miller-Urey experiments have also yielded, for example, hydrogen cyanide or formaldehyde or other precursors, it would be consequently justified to set up a second generation of simulation experiments using more or less concentrated solutions of pure hydrogen cyanide or formaldehyde or other precursors to produce a wealth of organic molecules. (For prebiotic chemistry of hydrogen cyanide see Schwartz et al. 1984; for the condensation of prebiotic formaldehyde see Gabel and Ponnamperuma 1967). These procedures must be criticized: their relevance to prebiotic conditions is extremely unlikely, because no mechanisms for the local accumulation and transient stabilization of such precursors can be supplied.

A summary of Miller's results is found in the textbook by Miller and Orgel (1974). His more recent work with a variety of less reducing atmospheres is described by Miller and Schlesinger (1984). A critical analysis of the evolutionary significance of these experiments has been published by Cairns-Smith (1982). Some constraints followed in prebiotic syntheses have been summarized by Orgel and Lohrmann (1974).

Also the "hot dilute soup" concept of the Oparin thesis has received heavy criticism (Brooks and Shaw 1978). The background of this criticism is immediately evident to any organic chemist: In an aqueous solution the reactive intermediates of the Miller-type experiments, e.g. aldehydes, amines, hydrogen cyanide and other compounds would readily interact to yield more complex and largely ill-defined materials, but practically no biologically relevant molecules. In view of this criticism and the poor production of biomolecules by simulation experiments, it is extremely unlikely that the first forms of life could have evolved spontaneously in a primordial soup.

Interactions of prebiotic organic materials with minerals have very likely played a decisive role in chemical evolution. For example, amino acids could have been selectively accumulated by the ion exchange properties of clays and other materials. Mineralic catalysts in an aqueous environment could have directed reactions more subtly and perhaps more favourably than an electric spark in a glass apparatus. The question of the involvement of minerals in chemical evolution needs to be explored more fully by experimentation, but it is definitely premature to postulate as an alternative a mineral origin of life (Cairns-Smith 1982).

The field "chemical evolution" is not as yet a mature scientific discipline. Many areas of research need to be developed more fully. In addition to the areas already mentioned, these include investigation of interstellar dust, comets and meteorites as sources of preformed organic compounds (Delsemme 1984; Agarwal et al. 1985; Anders and Owen 1977). A more detailed exploration of the planets and moons of our solar system, especially of the planet Mars, would also improve our knowledge about the possible conditions on the early Earth and thus help us to design more appropriate chemical evolution experiments. It appears now well established that the geological and climatological history of Mars and Earth was similar as far back as 4 to 4.6 billion years ago. But due to its smaller mass, Mars has lost most of its atmospheric N_2, CO_2 and H_2O by thermal diffusion and photochemical processes about 3 to 4 billion years ago and has cooled off much below the freezing point ever since. Therefore, no global erosion by water has changed the Martian surface during the last 3 to 4 billion years (Carr 1981; Squyres 1984).

D. The Origin of Genetic Information

I. Nucleic Acids First?

According to Muller (1966) the gene material is the initiator and the organizing basis of life. Ever since that date, until to-day, numerous experiments and concepts have been put forward to demonstrate the primacy of nucleotides and nucleic acids in the origin of life. Noteworthy milestones in the history of this field are in particular the experiments by Orgel and associates (see e.g. Inoue and Orgel 1983) on models of "self-replication" of RNA and the theoretical work by Eigen and Schuster on the origin of genetic information (the hypercyclus, e.g. see Eigen and Schuster 1979).

In order to promote their concept of nucleic acids first, some authors, e.g. Usher and Needels (1986), have even argued that since RNA may under certain circumstances show some biocatalytic activity (Bass and Cech 1984) protein-like enzymes were not required during the early evolution of life. Obviously these authors refuse to recognize that proteins are involved in every step of gene expression. These debates on primacies have not contributed much to the solution of any problem related to the origin of life. The model experiments and theories on the evolution of genetic information, primitive replication and translation mechanism may help to understand the evolution of the contemporary systems, but they do not explain their origins.

The literature on prebiotic syntheses with emphasis on possible origin of polynucleotides and their building blocks is vast, but inconclusive (as has already been pointed out in Sect. D), because no simulation experiment has yielded replicable polynucleotides, not even significant amounts of ribose or nucleosides. Some selected references are: Gabel and Ponnamperuma (1967); Fuller et al. (1972); Usher (1977), Schwartz (1981).

II. Clays First?

Genesis 2:6–7: "But there went up a mist from the earth, and watered the whole face of the ground. And the Lord God formed man of the dust of the ground, and breathed into his nostrils the breath of life, and man became a living soul."

Perhaps it would please both natural scientists and creational scientists if the ideas of Cairns-Smith (1982) on genetic takeover and the mineral origins of life could be verified. So far, however, his principal hypothesis is only a model without experimental foundation. It has been stated earlier in Section D, however, that the possible role of minerals in early prebiotic evolution, e.g. as catalyst, ion exchange materials or even as organizing matrices or structural entities needs to be explored more fully. The idea that the first forms of "life" were basically mineralic and not organic, however, provides no explanation for the origin of organic life.

III. (Proto)proteins First?

It has been demonstrated in Section D that at least certain types of amino acids are produced with substantial yields (a few %) in simulated prebiotic synthesis experiments. In order to form relatively pure polymers (protoproteins) from prebiotic amino acids these would have to be accumulated and purified by geophysical processes. Because of the good solubility of most amino acids in water and their electrochemical properties, plausible geological accumulation mechanisms can be seen in weathering of prebiotic organic materials, separation of the dissolved amino acids from other solutes by the ion exchange properties of minerals (e.g. zeolites or montmorillonites) and final concentration by evaporation. But the actual operation of such mechanisms under geological conditions has been established neither for amino acids nor for other prebiotic compounds. This is a weakness common to all "second generation" simulation experiments in which highly purified materials (e.g. amino acids, formaldehyde, HCN) that are actually only produced with modest yields in Miller-Urey-type experiments are used to demonstrate their prebiotic conversion into more complex biomolecules (e.g., peptides, ribose, adenine, nucleosides).

However, the geochemical stability of amino acids as demonstrated by their occurrence in Precambrian sediments and their chemical stability (see the reviews by Miller and Orgel 1974; Dose and Rauchfuss 1975 or Fox and Dose 1977) in connection with the significant yields of their formation in prebiotic simulation experiments suggests that prebiotic amino acids were privileged candidates for further condensation reactions.

In dilute aqueous solutions the chances for a condensation of amino acids to peptides, proteins or related materials are rather small for thermodynamic reasons. In order to synthesize polypeptides from amino acids, a condensation reaction between the reactant amino acids under elimination of water must take place.

In the laboratory the chemist usually "activates" the amino acids by converting them into more energy-rich derivates. In the living cell this activation is achieved by a primary reaction of amino acids with adenosine triphosphate. In any event, the synthesis of a peptide requires the removal of water and the addition of free energy according to the following equation:

$$H_3N^{\pm}CH(R)-COO^- + H_3N^{\pm}CH(R)-COO^- \Longrightarrow$$
$$H_3N^{\pm}CH(R)-CO-NH-CH(R)-COO^- + H_2O$$
$$\triangle G^\circ = 2\text{--}4 \text{ kcal/Mol.}$$

The amount of free energy, $? \triangle G$, in the formation of higher peptides is somewhat lower because the energy required to increase the distance between the H_3N^+ group (at the N-terminal end of the peptide) and the -COO⁻ group (at the N-terminal end) is lowered with the increasing length of the peptide chain. But the overall equilibrium remains so unfavourable that solutions 1 M in each amino acid of 20 different types would yield at equilibrium a 10^{-99} M concentration of polypeptides with a molecular weight of 12,000.

The thermodynamic barrier is surmountable by removal of the water formed as a by-product when peptide bonds are synthesized. This can be achieved by thermal evaporation of the water (usually at temperatures above 100°C) or by binding the water chemically with the help of condensing agents: In the prebiotic realm minerals or cyanamide could have acted as condensing agents. Areas with temperatures of 100°–180°C that are favourable for the thermal condensation were probably more frequent on the prebiotic Earth than they are now. Laboratory experiments have shown that thermal condensation of amino acids at these temperatures is by far more efficient than the processes involving chemical binding of the water released during condensation (Fox and Dose 1977). The disadvantage of the thermal polymerization, from the view point of contemporary science, however, rests in its complex chemistry and in the production of polymers whose structures are extremely difficult to analyze (see below for more details).

In particular, S.W. Fox and his associates (see the review in: Fox and Dose 1977) have shown that mixtures of amino acids can be polymerized by heat (80°C to 200°C) to yield protein-like non-random macromolecules (proteinoids). It has been demonstrated that the sequence of building blocks during thermal condensation is controlled by the chemical properties of the reacting amino acids and that their reactivity is in turn also influenced by their environment [pH, temperature, concentration, presence of minerals and other reactive components includiong the growing polymer (see Fox and Dose 1977)]. These experiments have led to a better understanding of some of the extremely complex reactions involved in the thermal condensation process. Very pronounced is the self-sequencing effect of glutamic acid, since it readily yields pyroglutamic acid which in turn is preferably found at the N-terminus of the growing peptide chain (Nakashima et al. 1977; Hartmann et al. 1981). It has also been shown that the partial "decomposition" of amino acids during the thermal condensation process yields new compounds of prebiotic significance which are partially incorporated into the polymer. Among these are, for example, flavines and pterines (Heinz et al. 1979). Thermal amino acid polymers (proteinoids) are probably no pure polypeptides, but heteropolymers with alternate sections of oligopeptide chains and chromophores (Hartmann et al. 1981) as the repetitive units. In view of such structures, the failure to sequence proteinoids by Edman degradation or with the help of peptidases becomes

plausible and the weak enzyme-like activities of some thermal proteinoids in oxidative deamination and reductive aminations become better explainable on the basis of
the presence of flavins and related groups (Krampitz et al. 1967, 1968).

Because the study of selective interactions of amino acids during the thermal
polymerization process is seriously perturbed by many side reactions, also prebiotic
model investigations on carbodiimide-mediated amino acid condensation in aqueous
solution have been made (Kenyon and Steinman 1969; Hartmann et al. 1984; Dose
et al. 1982). The carbodiimide of our choice was N-cyclohexyl-N'-[β -(methylmor-
pholino) -ethyl] carbodiimide p-toluenesulphonate (CME). This carbodiimide has
also found some application in organic chemical peptide synthesis (see Mutter 1979).
Typical "prebiotic condensing agents", e. g. HCN derivatives, are not suitable for detailed studies on sequence selection because of their involvement in complex side
reactions.

The CME-mediated copolymerization of dissolved amino acid mixtures containing
substantial proportions (above 10%) of glutamic acid or pyroglutamic acid usually
leads to short pyroglutamyl peptides, because the fastest reaction is the conversion of
glutamic acid to pyroglutamic acid, followed by the reaction of activated pyroglutamic
acid with the amino group of the next amino acid. In the absence of glutamic acid or
pyroglutamic acid the formation of longer chains with ten and more amino acids is
favoured. In principal, amino acids with short side chains like Ala or Gly are incorporated faster than those with bulky side chains like Leu, Val or Phe.

A copolymerization of an equimolar amino acid solution will therefore ideally yield
peptides with the more reactive amino acids at the N-terminus (e. g. N-terminal sequences like pyroglutamyl-glycin or glycyl-alanin). In the experiments described here
CME was applied in excess. Thus all carboxylic groups were activated. The amino
group of a carbodiimide-activated amino acid is more reactive than that of a free
amino acid. The kinetics of interactions of carboxyl-activated peptide chains with carboxyl-activated amino acids, however, is at present not easily explained because of
the lack of data. So far only the formation of various oligopeptides could be evaluated
and confirmed with the aid of computer simulation programmes (Hartmann et al.
1984).

Three-dimensional histogram plots of the sequence simulation of the copolymerization of pyroglutamic acid (P), alanine (A) and leucine (L) are shown in Fig. 1 (formation of pyroglutamyl amino acids and pyroglutamyl-dipeptides) and Fig. 2 (formation
of pyroglutamyl-tripeptides). The distribution of the peptides is a function of the reaction time (number of random steps), the reactivity of all reactants and their concentrations. The more reactive a component is, the faster will its concentration decrease (in
a closed system) and then the reactions of the less reactive components will prevail as
long as their concentrations are sufficiently high.

Sequence selection primarily occurs due to the reactivity of the amino group of the
next incoming amino acid (CME-activated, although its carboxylic group does not
react at this stage).

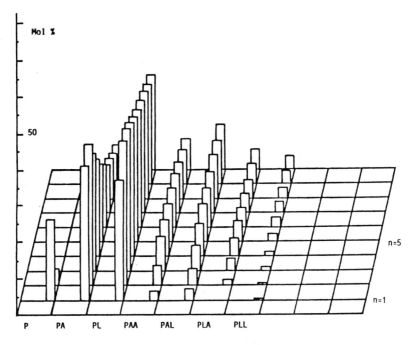

Fig. 1. Sequence simulation of the copolymerization of pyroglutamic acid *(P)*, alanine *(A)* and leucine *(L)* to dimers and trimers. (*n* number of random steps; 1 unit = 200 steps)

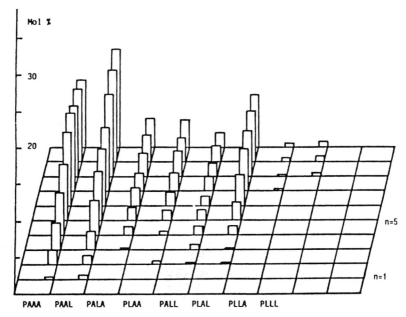

Fig. 2. Sequence simulation of the copolymerization of pyroglutamic acid *(P)*, alanine *(A)* and leucine *(L)* to tetramers. (*n* number of random steps; 1 unit = 1000 steps)

The relative reactivities of the individual amino acids vary from 1.0 (for glycine or alanine) to about 0.3 (for those with bulky side chains such as tryptophan, leucine, isoleucine or valine). It is interesting to note that similar relative reaction constants have been determined by Mutter, although in an entirely different context (Mutter 1979). While using the same carbodiimide (CME) in an aqueous system, Mutter had studied the relative reactivity of the carboxylic group in isobutyl-oxycarbonyl-pro-tected amino acids towards the amino group of alanine that was bound to polyethylene glycol.

The differences in the relative activities may appear too small to allow for a suffi-cient sequence selection, for example, to produce prebiotic polypeptides of specific structures and functions as required for protoenzymes. Moreover, it is not possible to evaluate to what extent the different abundances of the individual prebiotic amino acids have additionally contributed to the process of sequence selection, because the actual conditions of the prebiotic scene are not known. Another difficulty is that it is as yet almost impossible to design and to conduct a condensation experiment so that polypeptides with specific amino acid sequences are formed, that in turn yield spatial structures of desirable activity (e. g. of an RNA polymerase). At present, there is also not enough progress in the understanding of protein folding to solve the second diffi-culty even to a first approximation (Kolata 1986).

In principal, however, it appears not impossible that one family of protoproteins exhibited a kind of primitive nucleotide polymerase activity, an activity regarded as essential for the formation of the first replicable polynucleotides. But by prebiotic simulation experiments so far only very weak hydrolase, lyase or oxidoreductase ac-tivities of thermal polyamino acids (proteinoids) have been confirmed (Fox and Dose 1977). There is only one single report on a weak adenine nucleotide oligomerase activ-ity (AMP dimers and trimers formed from ATP) of proteinoid microspheres (Jungck and Fox 1973). The difficulties in designing the most appropriate simulation experi-ments add here to the known difficulties already mentioned: there are no satisfactory mechanisms to explain the origin of sufficient amounts of prebiotic nucleotides and the origin of chiral selection, an obligatory requirement for the formation of replica-ble polynucleotides.

If a plausible mechanism for the appearance of optically pure nucleotides on the prebiotic Earth could be presented, then it would be somewhat easier to demonstrate how the *first* replicable nucleic acids arose by protoprotein-controlled processes, be-cause in contemporary biological systems we find not only enzymes that catalyze the polymerization of nucleotides in the absences of a nucleic acid template; there are even some enzymes that — within certain limits — directly control the nucleotide sequence of their product.

Well-known examples are the addition of a CCA sequence to the 3'end of some pre-tRNA species, the addition of a poly(A) sequence to eukaryotic pre-mRNA or the ad-dition of the "cap"-group to the 5' end of pre-mRNA.

Most spectacular is the activity of the replicase of the *E. coli* phage Qβ. This en-zyme exhibits also an RNA polymerase activity which enables it to perform a de novo synthesis of RNA in the absence of a template in test tube experiments (Biebricher et al. 1986). Probably the enzyme begins with the synthesis of a variety of small oligonucleotides which then are linked together to yield a population of different polynucleotides. Most of these cannot be replicated, because they are not recognized by the replicase site of the enzyme. But it appears that those RNA's that contain a 3'-terminal CCC sequence and another intramolecular CCC sequence at a defined distance are at least slowly-replicated (Küppers 1979). The enzyme will first synthesize de novo polynucleotide sequences and then pick those most suitable for replication by trial and selection in a template-free process (Biebricher et al. 1986). It has been shown that these RNA's already have a high secondary structure (involving complementary nucleotide sequences). These experiments therefore principally establish that ordered RNA molecules can be formed in a template-free protein-directed synthesis.

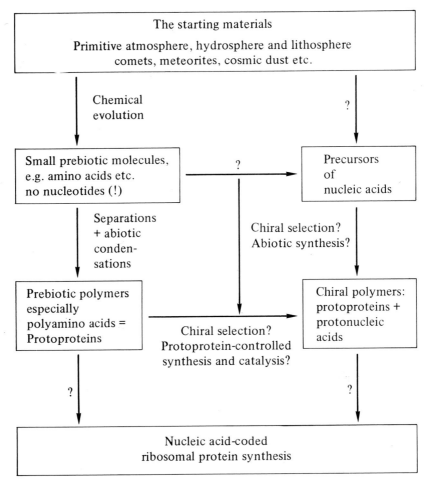

Fig. 3. The origin of nucleic acid-coded ribosomal protein synthesis is unknown

However, even if protoproteins with Qβ replicase-like activities could be produced by prebiotic simulation experiments, many more protoprotein and nucleic acid species would be required for the assembly of a first nucleic acid-coded protein biosynthesis system. The difficulties that must be overcome are at present beyond our imagination.

E. Conclusions

So far we are almost completely ignorant with respect to the conditions on the prebiotic Earth. In view of the experimental data available it appears possible that polypeptides were the first informative molecules. But it is at present impossible to re-construct prebiotic pathways in order to firmly establish the primacy of protoproteins or to show how replicative polynucleotides or other chiral systems arose.

The scheme shown in Fig. 3 is a scheme of ignorance. Without fundamentally new insights in evolutionary processes, perhaps involving new modes of thinking, this ignorance is likely to persist.

References

Anders E (1977) Mars and Earth: origin and abundance of volatiles. Science 198:453–465
Argarwal VK, Schutte W, Greenberg JM et al. (1985) Photochemical reactions in interstellar grains-photolysis of CO, NH$_3$ and H$_2$O. Origins Life 16:21–40
Bass BL, Cech TR (1984) Specific interaction between the self-splicing of RNA of Tetrahymena and its guanosine substrate: implication for biological catalysis by RNA. Nature 308:820–826
Biebricher CK, Eigen M, Luce R (1986) Template-free RNA synthesis by Qβ replicase. Nature 321:89–91
Brooks J, Shaw G (1978) A critical assessment of the origin of life. Origins Life 9:597–606
Cairns-Smith AG (1982) Genetic takeover and the mineral origins of life. Cambridge University Press, New York
Calvin M (1969) Abiogenic information coupling between nucleic acid and protein or, how protein and DNA were married. Proc Roy Soc Edinburgh, Sect B70 part 4, 273–294
Carr MH (1981) The surface of Mars. Yale University Press
Crick FHC (1968) The origin of the genetic code, J Mol Biol 38:367–379
Crick FHC, Orgel L (1973) Directed panspermia. Icarus 19:341–346
Darwin C (1859) The origin of species, Murray, London (reprinted: New American Library, New York, 1958)
Darwin C (1959) Some unpublished letters (Sir Gavin de Beer, ed). Notes Res R Soc Lond 14:1
Darwin F (1896) The life and letters of Charles Darwin, Vol II. Appleton, New York, p 202
Delsemme AH (1984) The cometary connection with prebiotic chemistry. Origins Life 14:51–60
Dose K, Rauchfuss H (1975) Chemische Evolution und der Ursprung lebender Systeme. Wissenschaftliche Verlagsgesellschaft, Stuttgart
Dose K, Hartmann J, Brand CM (1982) Formation of specific amino acid sequences during carbodiimide-mediated condensation of amino acids in aqueous solution. Biosystems 15:195–200

Eigen M, Schuster P (1979) The hypercycle. Springer, Berlin, Heidelberg, New York

Farley J (1977) The spontaneous generation controversy from Descartes to Oparin. Hopkins University Press, Baltimore

Folsome CE (1979) The origin of life. Freeman, San Francisco

Fox SW, Dose K (1977) Molecular evolution and the origin of life. Dekker, New York

Fuller WD, Sanchez RA, Orgel LE (1972) Studies in prebiotic synthesis, VII. Solid state synthesis of purine nucleosides. J Mol Evol 1:249–257

Gabel NW, Ponnamperuma C (1967) Model for the origin of monosaccharides. Nature 216:453–455

Haeckel E (1868) Natürliche Schöpfungsgeschichte. Reimer, Berlin

Hartmann J, Brand CM, Dose K (1981) Formation of specific amino acid sequences during thermal polymerization of amino acids. Biosystems 13:141–147

Hartmann J, Nawroth TH, Dose K (1984) Formation of specific amino acid sequences during carbodiimide-mediated condensation of amino acids in aqueous solution, and computer-simulated sequence generation. Origins Life 14:213–220

Heinz B, Ried W, Dose K (1979) Thermische Erzeugung von Pterinen und Flavinen aus Aminosäuregemischen. Angew Chem 91:511–517

Inue T, Orgel L (1983) A nonenzymatic RNA polymerase model. Science 219:859–862

Jungck JR, Fox SW (1973) Synthesis of oligonucleotides by proteinoid microspheres. Naturwissenschaften 60:425–427

Kenyon DH, Steinman G (1969) Biochemical predestination. McGraw-Hill, New York

Kolata G (1986) Protein folding is now a challenge that cannot be ignored. Science 233:1038–1039

Krampitz G, Diel S, Nakashima T (1967) Aminotransferase Aktivität von Polyanhydro-α-aminosäuren (Proteionoiden). Naturwissenschaften 54:516–517

Krampitz G, Baars-Diel S, Haas W, Nakashima T (1968) Aminotransferase activity of thermal polylysine. Experientia (Basel) 24:140–142

Küppers B (1979) Towards an experimental analysis of molecular self-organization and precellular Darwinian evolution. Naturwissenschaften 66:228–243

Lawless JG, Peterson E (1976) Amino acids in carbonaceous chondrites. Origins Life 6:3–8

Miller SL (1953) A production of amino acids under possible primitive earth conditions. Science 117:528–529

Miller SL, Orgel LE (1974) The origins of life on earth. Englewood Cliffs, New York, Prentice-Hall

Miller SL, Schlesinger G (1984) Carbon and energy yields in prebiotic syntheses using atmospheres containing CH_4, CO and CO_2. Origins Life 14:83–90

Miller SL, Urey HC, Oro J (1976) Origin of organic compounds on the primitive earth and in meteorites. J Mol Evol 9:59

Muller HJ (1966) The gene material as the initiator and the organizing basis of life. Am Nat 100:493–517

Mutter M (1979) Studies on the coupling rates in liquid-phase peptide synthesis using competition experiments. Int J Pept Protein Res 13:274–277

Nakashima T, Jungck JR, Fox SW, Lederer E, Das BC (1977) A test for randomness in peptides isolated from a thermal polyamino acid. Int J Quantum Chem Quantu Biol Symp 4:65–72

Oparin AI (1924) The origin of life, 1st edn. (Russian: Proiskhozhdenie Zhizny), Moscow: Rabochii

Oparin AI (1957) The origin of life on Earth, 3rd. edn. Academic Press, London

Oparin AI (1964) Life, its nature, origin and development (translated from the Russian by A. Synge). Academic Press, London

Orgel L, Lohrmann R (1974) Prebiotic chemistry and nucleic acid replication. Accounts Chem Res 7:368–377

Pasteur L (posthumous) (1922) In: Vallery-Radot P (ed) Oeuvres de Pasteur, vol. II. Masson, Paris, pp 210–294, 328–346

Redi F (posthumous) (1909) Experiments on the generation of insects, (translated by M. Bigelow). The Open Court, Lasalle IL

Richter H (1865) Zur Darwinschen Lehre. Schmidts Jahrb Gesamte Med 126:243–249

Schwartz AW (1981) Chemical evolution — the genesis of the first organic compounds. In: Duursma EK, Dawson RD (eds) Marine organic chemistry. Elsevier, Amsterdam, pp 7–30

Schwartz AW, Voet AB, van der Veen M (1984) Recent progress in the prebiotic chemistry of HCN. Origins Life 14:91–98

Shapiro R (1984) The improbability of prebiotic nucleic acid synthesis. Origins Life 14:565–570

Shapiro R (1986a) Prebiotic ribose synthesis: a critical analysis. Abstracts of the 5th ISSOL Meeting, Berkeley, CA: ISSOL, pp 107–108

Shapiro R (1986b) Origins. Summit Books, New York

Squyres SW (1984) The history of water on Mars. Annu Rev Earth Planet Sci 12:83–106

Thaxton CB, Bradley WL, Olsen R (1984) The mystery of life's origin: Reassessing current theories. Philosophical Library, New York

Usher DA (1977) Early chemical evolution of nucleic acids: A theoretical model. Science 196:311–313

Usher DA, Needels MC (1986) Translation models. Abstracts of the 5th ISSOL Meeting, Berkeley ISSOL

Subject Index

114